Spectroscopic methods in organic chemistry

Fourth edition
Revised

Spectroscopic methods in organic chemistry

Fourth edition
Revised

Dudley H. Williams, MA, PhD, ScD, FRS
Fellow of Churchill College, Cambridge

Ian Fleming, MA, PhD, ScD
Fellow of Pembroke College, Cambridge

McGRAW-HILL BOOK COMPANY

London · New York · St Louis · San Francisco · Auckland · Bogotá · Caracas
Hamburg · Lisbon · Madrid · Mexico · Milan · Montreal · New Delhi · Panama · Paris
San Juan · São Paulo · Singapore · Sydney · Tokyo · Toronto

Published by
McGRAW-HILL Book Company (UK) Limited
Shoppenhangers Road
Maidenhead · Berkshire · England
Telephone Maidenhead (0628) 23432
Cables McGRAW-HILL MAIDENHEAD Telex 848484
Fax 0628 770224

British Library Cataloguing in Publication Data
Williams, Dudley H. (Dudley Howard), 1937–
 Spectroscopic methods in organic
 chemistry.—4th ed., rev.
 1. Organic compounds. Spectroscopy
 I. Title II. Fleming, Ian, 1935–
 547.3′0858
 ISBN 0–07–707212–X

Library of Congress Cataloging-in-Publication Data
Williams, Dudley H.
 Spectroscopic methods in organic
 chemistry/Dudley H. Williams, Ian Fleming.—4th ed., rev.
  p. cm.
 Includes bibliographies and index.
 ISBN 0–07–707212–X

 1. Chemistry, Organic. 2. Spectrum analysis.
 I. Fleming, Ian, 1935–   . II. Title.
 QD272.S6W54   1989
 547.3′0858–dc20   89–7991

34 WC 9210

Printed and bound in Great Britain by William Clowes Limited, Beccles and London

Contents

Preface   vii

Preface to fourth edition   ix

Foreword   xi

1   Ultraviolet and visible spectra   1

2   Infrared spectra   29

3   Nuclear magnetic resonance spectra   63

4   Mass spectra   150

5   Structure elucidation by joint application of UV, IR, NMR,
         and mass spectroscopy   199

Answers to problems   258

Index   259

Preface

We have written this book as a guide to the interpretation of the ultraviolet, infrared, nuclear magnetic resonance and mass spectra of organic compounds; it is intended both as a textbook suitable for a first course in the subject and as a handbook for practising organic chemists.

Spectroscopic methods are now used at some point in the solution of almost all problems in organic chemistry. Three of these methods rely on the selective absorption of electromagnetic radiation by organic molecules. The first method, ultraviolet spectroscopy, is used to detect conjugated systems, because the promotion of electrons from the ground state to the excited state of such systems gives rise to absorption in this region. The second, infrared spectroscopy, is used to detect and identify the vibrations of molecules, and especially the characteristic vibrations of the double and triple bonds present in many functional groups. The third method, nuclear magnetic resonance spectroscopy, uses a longer wavelength of the electromagnetic spectrum to detect changes in the alignment of nuclear magnets in strong magnetic fields. Absorption is observed from such nuclei as ^1H, ^{13}C, ^{15}N, ^{19}F, and ^{31}P; and the precise frequency of absorption is a very sensitive measure of the magnetic, and hence the chemical, environment of such nuclei. Moreover, the number and disposition of neighbouring magnetic nuclei influence the appearance of that absorption in a well-defined way. The result, particularly with the ubiquitous hydrogen and ^{13}C nuclei, is a considerable gain in information about the arrangement of functional groups and hydrocarbon residues in a molecule. The fourth method, mass spectrometry, measures the mass-to-charge ratio of organic ions created by electron bombardment. Structural information comes from the moderately predictable fragmentation organic molecules undergo; the masses of the fragment ions can often be related to likely structures.

Other regions of the electromagnetic spectrum are often used to determine the structure of organic molecules. X-ray diffraction can be used to pinpoint centres of high electron density (that is, the atoms). Microwave absorption is used to measure molecular rotations. Electron spin resonance, also using radio frequency signals, detects unpaired electrons, and can be used to measure the distribution of electron density in radicals. Optical rotatory dispersion and circular dichroism, using visible and ultraviolet light, measure the change in rotatory power of molecules, as the wavelength of the polarized light is changed; such measurements can often be related to the absolute configuration of molecules. Other physical methods, such as the measurement of pKs, reaction rates and dipole moments, are also used by organic chemists for structure determination. But all of these methods are more specialized than the four spectroscopic methods described in this book: these four methods are so regularly used that all organic chemists need to know about them.

We have kept discussion of the theoretical background to a minimum, since correlations between spectra and structure can successfully be made without detailed theoretical knowledge; this aspect of the subject has, moreover, been covered in many books, including the book by C. N. Banwell, *Fundamentals of Molecular Spectroscopy*. We have instead discussed in each chapter the kind of information given by each of the four spectroscopic methods, and we have described how to read each kind of spectrum to get that information out of it. We have included in each chapter tables of values: UV maxima, IR frequencies, NMR chemical shifts and coupling constants, and common mass fragments found in mass spectra—all of which are regularly needed for the day-to-day interpretation of spectra.

Finally, in Chapter 5, we give four examples of the way in which the spectroscopic methods can be brought together to solve fairly simple structural problems. There are then 18 sets of spectra at the end of the chapter. These are intended to supplement an organized course which the student is attending. Throughout the book we have stressed the application of spectroscopic methods to structure determination, though the application to other problems is limited only by the ingenuity of the researcher and analyst.

Ian Fleming
Dudley H. Williams
Cambridge

Preface to Fourth Edition

In preparing a fourth edition, we have completely rewritten chapters 3 (NMR) and 4 (Mass Spectra), and we have made a few small additions to the other chapters. In the chapter on NMR spectroscopy, we have integrated, enlarged, and clarified the discussion of ^{13}C and ^1H NMR spectra, dealing first with the chemical shift of both nuclei, and continuing with their coupling properties. We have added sections on new methods (difference spectra, COSY, NOESY, and their like), keeping the discussion to the minimum needed by a student who wants to know only what these techniques can do and what the spectra produced by each look like. We have also reorganized the tables of data collected at the end of the chapter and added substantially to them, to make the book more lastingly useful to those students who continue with organic chemistry later into their careers. We thank Dr Jeremy Sanders for his help with this chapter.

In the chapter on mass spectroscopy, we have increased the coverage of the newer methods of producing ions (CI, FD, plasma desorption, and FAB). Additionally, we have emphasized that a mass spectrum is largely determined by the energetics of unimolecular reactions and, unlike the other methods covered in this book, is not a true spectroscopic technique. These principles will allow the student to have a better feel about what to expect in a mass spectrum; we hope that this improves upon the traditional empirical approach to the subject. We have added sections which show that larger polar molecules can now be handled, either by particle ionization, or by HPLC/MS combinations.

Cambridge

Ian Fleming
Dudley H. Williams

Foreword

Spectroscopy is the cornerstone of the organic chemistry of the second half of the 20th century. The introduction of ultraviolet spectroscopy (UV) in the 1930s and infrared spectroscopy (IR) in the 1940s provided chemists with effective methods for the recognition of functionality in organic molecules. For the first time, structural information could be obtained by small-scale, non-destructive experiments. The revolution continued with the introduction of mass spectrometry (MS) in the 1950s. This experiment provided the molecular formula for a compound, and gave an insight into structure from the fragmentation pattern.

However, the analytical method that has had the greatest impact on science has been nuclear magnetic resonance spectroscopy (NMR). The effect of NMR has come in three distinct waves. Although the first applications in organic chemistry came in the 1950s, the tool was not widely used until the advent of the Varian Associates A-60 spectrometer in the early 1960s. The new experiment provided the final piece of the structural puzzle in many cases; UV and IR gave the functionality, MS gave the formula, and NMR and MS together allowed one to put together the molecular skeleton. The effect on organic chemistry was immediate and electrifying; by the end of that decade, virtually every publication dealing with organic chemistry included NMR data as the most important structural evidence. The pace of progress in structure determination increased perceptibly.

A second identifiable stage in the introduction of NMR into structural chemistry resulted from application of the fourier transform (FT) method to the nuclear magnetic resonance experiment. The development of the FT instrument in the 1970s had an effect on organic chemistry and biological chemistry almost as far-reaching as the initial introduction of proton NMR a decade earlier. The application of the FT technique to proton NMR allowed the use of this important method with surprisingly small samples. It also solved the dual problems of the low natural abundance and relatively small magnetic sensitivity of carbon-13, and allowed the NMR experiment to be applied to this isotope. With carbon-13 NMR (CMR) the chemist could look directly at the carbon backbone of molecules; when CMR was used in combination with proton NMR and the new separation techniques (gas chromatography, thin layer chromatography, high performance liquid chromatography) structural elucidation took on new dimensions.

The final surge of activity has come in the present decade, and is a daughter of the computer revolution. 'Two-dimensional' (2D) NMR is a still-burgeoning aspect of the method that may influence the way chemists think about structure determination more than any of the earlier spectroscopic techniques. With this powerful family of experiments, one may correlate proton and carbon spectra, trace the connectivity of a molecular skeleton, and even determine non-bonded distances within molecules and between different molecules. It is conceivable that 2D NMR will eventually permit the kind of detailed structural 'photograph' that can now be obtained only in the crystalline state by the costly and time-consuming technique of x-ray crystallography.

This book, the Fourth Edition of a classic work on spectroscopy, concentrates on the practical aspect of *using* spectroscopic techniques to solve structural problems. It is written at a level that is suitable for an advanced undergraduate or graduate course in applied spectroscopy, but will also give practicing chemists a valuable overview of the subject, as well as a good introduction to newer techniques (2D NMR and recently-introduced methods of producing ions for mass spectrometry). It is a resource that should be on the desks of *all* graduate students beginning organic chemistry, and could be read with profit by many of their professors as well.

Clayton H. Heathcock
Berkeley
November, 1986

1. Ultraviolet and visible spectra

1.1 *Introduction.* 1.2 *The energy of electronic excitation.* 1.3 *The absorption laws.* 1.4 *Measurement of the spectrum.* 1.5 *Vibrational fine structure.* 1.6 *Choice of solvent.* 1.7 *Selection rules and intensity.* 1.8 *Chromophores.* 1.9 *Solvent effects.* 1.10 *Searching for a chromophore.* 1.11 *Standard works of reference.* 1.12 *Definitions.* 1.13 *Conjugated dienes.* 1.14 *Polyenes.* 1.15 *Polyeneynes and poly-ynes.* 1.16 *Ketones and aldehydes; $\pi \rightarrow \pi^*$ transitions.* 1.17 *Ketones and aldehydes; $n \rightarrow \pi^*$ transitions.* 1.18 *αβ-unsaturated acids, esters, nitriles and amides.* 1.19 *The benzene ring.* 1.20 *Substituted benzene rings.* 1.21 *Polycyclic aromatic hydrocarbons.* 1.22 *Heteroaromatic compounds.* 1.23 *Quinones.* 1.24 *Porphyrins, chlorins and corroles.* 1.25 *Non-conjugated interacting chromophores.* 1.26 *The effect of steric hindrance to coplanarity. Bibliography.*

1.1 Introduction

The visible and ultraviolet spectra of organic compounds are associated with transitions between electronic energy levels. The transitions are generally between a bonding or lone-pair orbital and an unfilled non-bonding or anti-bonding orbital. The wavelength of the absorption is then a measure of the separation of the energy levels of the orbitals concerned. The highest energy separation is found when electrons in σ-bonds are excited, giving rise to absorption in the 120–200 nm (1 nm = 10^{-7} cm = 10 Å = 1 mμ) range. This range, known as the vacuum ultraviolet, since air must be excluded from the instrument, is both difficult to measure and relatively uninformative. Above 200 nm, however, excitation of electrons from p- and d-orbitals and π-orbitals, and, particularly, π-conjugated systems, gives rise to readily measured and informative spectra.

1.2 The energy of electronic excitation

The energy is related to wavelength by Eq. 1.1.

$$E(\text{kJ mol}^{-1}) = \frac{1.19 \times 10^5}{\lambda(\text{nm})} \tag{1.1}$$

Thus 297 nm, for example, is equivalent to 400 kJ (≈ 96 kcal)—enough energy, incidentally, to initiate many interesting reactions; compounds should not, therefore, be left in the ultraviolet beam any longer than is necessary.

1.3   The absorption laws

Two empirical laws have been formulated about the absorption intensity. *Lambert's law* states that the fraction of the incident light absorbed is independent of the intensity of the source. *Beer's law* states that the absorption is proportional to the number of absorbing molecules. From these laws, the remaining variables give the Eq. 1.2.

$$\log_{10} \frac{I_0}{I} = \varepsilon.l.c \tag{1.2}$$

I_0 and I are the intensities of the incident and transmitted light respectively, l is the path length of the absorbing solution in centimetres, and c is the concentration in moles/litre. $\log_{10}(I_0/I)$ is called the absorbance or optical density; ε is known as the molar extinction coefficient and has units of $1000 \text{ cm}^2 \text{ mol}^{-1}$ but the units are, by convention, never expressed.

1.4   Measurement of the spectrum

The ultraviolet or visible spectrum is usually taken of a very dilute solution. An appropriate quantity of the compound (often about 1 mg when the compound has a molecular weight of 100–200) is weighed accurately, dissolved in the solvent of choice (see below), and made up to, for instance, 100 ml. A portion of this is transferred to a silica cell. The cell is so made that the beam of light passes through a 1 cm thickness (the value l in Eq. 1.2) of solution. A matched cell containing pure solvent is also prepared, and each cell is placed in the appropriate place in the spectrometer. This is so arranged that two equal beams of ultraviolet or visible light are passed, one through the solution of the sample, one through the pure solvent. The intensities of the transmitted beams are then compared over the whole wavelength range of the instrument. In most spectrometers there are two sources, one of 'white' ultraviolet and one of white visible light, which have to be changed when a complete scan is required. Usually either the visible or ultraviolet alone is sufficient for the purpose in hand. The spectrum is plotted automatically on most machines as a $\log_{10}(I_0/I)$ ordinate and λ abscissa. For publication and comparisons these are often converted to an ε versus λ or $\log \varepsilon$ versus λ plot. The unit of λ is almost always nm. Strictly speaking the intensity of a transition is better measured by the area under the absorption peak (when plotted as ε versus frequency) than by the intensity of the maximum of the peak. For several reasons, most particularly convenience and the difficulty of dealing with overlapping bands, the latter procedure is adopted in everyday use. Spectra are quoted, therefore, in terms of λ_{max}, the wavelength of the absorption peak, and ε_{max}, the intensity of the absorption peak as defined by Eq. 1.2.

1.5   Vibrational fine structure

The excitation of electrons is accompanied by changes in the vibrational and rotational quantum numbers so that what would otherwise be an absorption *line* becomes a broad peak containing vibrational and rotational fine structure. Due to interactions of solute with solvent molecules this is usually blurred out, and a smooth curve is observed. In the vapour phase, in non-polar solvents, and with certain peaks (e.g. benzene with the 260 nm band), the vibrational fine structure is sometimes observed.

1.6 Choice of solvent

The solvent most commonly used is 95 per cent ethanol (commercial absolute ethanol contains residual benzene which absorbs in the ultraviolet). It is cheap, a good solvent, and transparent down to about 210 nm. Fine structure, if desired, may be revealed by using cyclohexane or other hydrocarbon solvents which, being less polar, have least interaction with the absorbing molecules. Table 1.1 gives a list of common solvents and the minimum wavelength from which they may be used in 1 cm cells.

Table 1.1 Some solvents used in ultraviolet spectroscopy

Solvent	Minimum wavelength for 1 cm cell, nm
Acetonitrile	190
Water	191
Cyclohexane	195
Hexane	201
Methanol	203
Ethanol	204
Ether	215
Methylene dichloride	220
Chloroform	237
Carbon tetrachloride	257

The effect of solvent polarity on the position of maxima is discussed in Sec. 1.9.

1.7 Selection rules and intensity

The irradiation of organic compounds may or may not give rise to excitation of electrons from one orbital (usually a lone-pair or bonding orbital) to another orbital (usually a non-bonding or anti-bonding orbital). It can be shown that:

$$\varepsilon = 0.87 \times 10^{20} \, P . \mathbf{a} \tag{1.3}$$

where P is called the transition probability (with values from 0 to 1) and $\mathbf{a}$ is the target area of the absorbing system; the absorbing system is usually called a chromophore. With common chromophores of the order of 10 Å long, a transition of unit probability will have an ε value of 10^5. This is close to the highest observed values, though—with unusually long chromophores—values in excess of this have been measured. In practice, a chromophore giving rise to absorption by a fully allowed transition will have ε values greater than about 10 000, while those with low transition probabilities will have ε values below 1000. The important point is that, in general, *the longer a particular kind of chromophore, the more intense the absorption*.

There are many factors which affect the transition probability of any particular transition. In the first place there are rules about which transitions are allowed and which are forbidden. These are complicated because they are a function of the symmetry and multiplicity both of the ground state and excited state orbitals concerned. The spectra of chromophores, with $\varepsilon_{\mathrm{max}}$ less than about 10 000, are the result of 'forbidden' transitions. Two very important and 'forbidden' transitions are observed: the $n \rightarrow \pi^*$ band near 300 nm of ketones, with ε values of the order of 10 to 100; and the benzene 260 nm band and its equivalent in more complicated systems, with ε values from 100 upwards. Both occur because the symmetry which makes absorption strictly forbidden is broken up by molecular vibrations and—in the latter case—by substitution. Both types are discussed further under the sections on ketones and aromatic systems.

In this and the following discussions a very simplified theoretical picture is given; there is considerable danger in being satisfied with so little in so well developed a subject. The books by Jaffé and Orchin and by Murrell, listed in the bibliography, give excellent accounts of the state of the art.

1.8   Chromophores

The word chromophore is used to describe the system containing the electrons responsible for the absorption in question. Most of the simple unconjugated chromophores described in Table 1.2 below give rise to such high-energy, and therefore such short-wavelength absorption, that they are of little use.

Table 1.2   The absorption of simple unconjugated chromophores

Chromophore	Transition notation†	λ_{max}, nm
σ-bonded electrons		
$>$C—C$<$ and $>$C—H	$\sigma \rightarrow \sigma^*$	~ 150
Lone-pair electrons		
—Ö—	$n \rightarrow \sigma^*$	~ 185
—N̈$<$	$n \rightarrow \sigma^*$	~ 195
—S̈—	$n \rightarrow \sigma^*$	~ 195
$>$C=Ö	$n \rightarrow \pi^*$	~ 300
$>$C=Ö	$n \rightarrow \sigma^*$	~ 190
π-bonded electrons		
$>$C=C$<$ (isolated)	$\pi \rightarrow \pi^*$	~ 190

† There are many other notations used.

One of the few useful simple unconjugated chromophores is the very weak forbidden n → π* transition of ketones mentioned earlier which appears in the 300 nm region and is of particular importance in connection with optical rotatory dispersion. This band is due to the excitation of one of the lone pair of electrons (designated n) on the oxygen atom to the lowest anti-bonding orbital (designated π*) of the carbonyl group. It is discussed further in the sections on solvent effects and on ketones.

The important chromophores are those in which conjugation is present. An isolated double bond or lone pair of electrons gives rise to a strong absorption maximum at about 190 nm, corresponding to the transition x in Fig. 1.1, at too short a wavelength for convenient measurement. When the molecular orbitals of two isolated double bonds are brought into conjugation, the energy level of the highest occupied orbital is raised and that of the lowest unoccupied anti-bonding orbital lowered (Fig. 1.1).

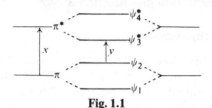

Fig. 1.1

The π → π* transition, which is occasioned by absorption, is now associated with the smaller value y. This transition appears in the spectrum of butadiene as a strong, easily detected, and easily measured maximum at 217 nm. The same principle governs the energy levels when unlike chromophores, e.g. those of an $\alpha\beta$-unsaturated ketone, are brought together. For instance, methyl vinyl ketone has an absorption maximum at 225 nm, while neither a carbonyl group nor an isolated double bond has a strong maximum above 200 nm.

When more than two π-bonding orbitals overlap, that is when the chromophore is a longer conjugated system, the separation of the energy levels is further reduced, and absorption occurs at longer wavelength. A long conjugated polyene, like carotene, absorbs, quite obviously since it is coloured, in the visible. The most important point to be made is that, in general, *the longer the conjugated system, the longer the wavelength of the absorption maximum*.

The rules and correlations possible with the spectra of conjugated dienes, $\alpha\beta$-unsaturated ketones and some substituted benzene ring compounds are given in Secs 1.13, 1.16, and 1.20. With complicated chromophores, predictions become more difficult. The usual procedure, when one is confronted with the ultraviolet spectrum of an unknown substance, is to compare the spectrum, in its general shape and in the intensity and position of its peaks, with the spectra of reasonable model compounds. These models are chosen to possess as nearly as possible the same chromophore as that suspected for the unknown.

1.9 Solvent effects

π → π*. The Frank–Condon principle states that during the electronic transition atoms do not move. Electrons, however, including those of the solvent molecules, may reorganize. Most transitions result in an excited state more polar than the ground state;† the dipole–dipole interactions with solvent molecules will, therefore, lower the energy of the excited state more than that of the ground state. Thus it is usually observed that ethanol solutions give longer wavelength maxima than do hexane solutions. In other words, there is a small red shift of the order of 10–20 nm in going from hexane as solvent to ethanol.

$n \rightarrow \pi^*$. The weak transition of the oxygen lone pair in ketones—the $n \rightarrow \pi^*$ transition—shows a solvent effect in the opposite direction. The solvent effect is now due to the lesser extent to which solvents can hydrogen bond to the carbonyl group in the excited state. In hexane solution, for example, the absorption maximum of acetone is at 279 nm ($\varepsilon = 15$), whereas in aqueous solution the maximum is at 264.5 nm. The shift in this direction is known as a blue shift.

1.10   Searching for a chromophore

There is no easy rule or set procedure for identifying a chromophore—too many factors affect the spectrum and the range of structures which can be found is too great. The examination of a spectrum with particular regard for the following points is the first step to be taken.

The complexity and the extent to which the spectrum encroaches on the visible region.   A spectrum with many bands stretching into the visible shows the presence of a long conjugated or a polycyclic aromatic chromophore. A compound giving a spectrum with only one band (or only a few bands) below about 300 nm probably contains only two or three conjugated units.

The intensity of the bands, particularly the principal maximum and the longest wavelength maximum.   This observation can be very informative. Simple conjugated chromophores such as dienes and $\alpha\beta$-unsaturated ketones have ε values of 10 000–20 000. The longer simple conjugated systems have principal maxima (usually also the longest wavelength maxima) with correspondingly higher ε values. Very low intensity absorption bands in the 270–350 nm region, on the other hand, with ε values of 10–100, are the result of the $n \rightarrow \pi^*$ transition of ketones. In between these extremes, the existence of absorption bands with ε values of 1000–10 000 almost always shows the presence of an aromatic system. Many unsubstituted aromatic systems show bands with intensities of this order of magnitude, the absorption being the result of a transition with a low transition probability, low because of the symmetry of the ground and excited states. When the aromatic nucleus is substituted with groups

---

† This transition is commonly visualized in valence bond terms with the ground state represented without charge separation and the excited state as the dipolar species.

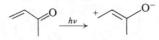

Such representations are over-simplified, and misleading: the dipolar structure is not *the* structure of the excited state, rather it is a more important contributor to the excited state than to the ground state. Since the valence bond technique is less exact and less revealing in this field than is the molecular orbital theory, the latter should be used on all occasions.

which can extend the chromophore, strong bands with ε values above 10 000 appear, but bands with ε values below 10 000 are often still present.

Having made these observations, one should search for a model system which contains the chromophore and therefore gives a similar spectrum to that which is being examined. This may be difficult in rare cases; but so many spectra are now known, and the changes caused by substitution so well documented, that the task can be a simple one. The first tool which an organic chemist requires is a general knowledge of the simple chromophores and the changes which structural variations make in the absorption pattern. Sections 1.13–1.26 give a very brief account of these topics. The remaining task, that of searching through the literature, is greatly facilitated by the existence of the indexes and compilations which are described in Sec. 1.11. The usefulness of these books will be greatly increased by a general knowledge of organic chemistry on which to base a guess as to what chromophores are likely to be known and in what compounds they may be found.

The search for a chromophore is also likely to be assisted by the other physical methods described in this book. The range of structures in which a search must be made can be narrowed, for example, to aromatic compounds on the strength of infrared or NMR aromatic C—H absorptions. Similarly the presence of an $\alpha\beta$-unsaturated ketone may be inferred from the C=O stretching vibration observed in the infrared spectrum and confirmed from the ultraviolet spectrum, and the extent of alkylation deduced by a consideration of Woodward's rules (Sec. 1.16) and by reference to the NMR spectrum. A very important stage in determining the structure of a natural product is the positive identification of the chromophore, by comparison of the spectrum with that of some known model compound.

1.11   Standard works of reference

In the search for a model chromophore, a number of source books are available. In addition, several textbooks, larger than the single chapter of this book, are devoted to the subject and are mentioned in the bibliography at the end of this chapter.

The major collection of *data* is *Organic Electronic Spectral Data*, Wiley, New York, Vols 1–21 (1960–1985). This most valuable collection has been prepared by a complete search of the major journals from 1945 to 1979. The compounds are indexed by their empirical formulae, and absorption maxima are quoted together with literature references.

Another substantial collection of spectra is the *Sadtler Handbook of Ultraviolet Spectra*, Heyden, London.

1.12   Definitions

The following words and symbols are commonly used.

Red shift or *bathochromic effect*.   A shift of an absorption maximum towards longer wavelength. It may be produced by a change of medium, or by the presence of an auxochrome.

Auxochrome. A substituent on a chromophore which leads to a red shift. For example, the conjugation of the lone pair on the nitrogen atom of an enamine has shifted the absorption maximum from the isolated double bond value of 190 nm to about 230 nm. The nitrogen substituent is the auxochrome. An auxochrome, then, extends a chromophore to give a new chromophore.

Blue shift or *hypsochromic effect.* A shift towards shorter wavelength. This may be caused by a change of medium and also by such phenomena as the removal of conjugation. For example, the conjugation of the lone pair of electrons on the nitrogen atom of aniline with the π-bond system of the benzene ring is removed on protonation. Aniline absorbs at 230 nm (ε 8600), but in acid solution the main peak is almost identical with that of benzene, being now at 203 nm (ε 7500). A blue shift has occurred.

Hypochromic effect. An effect leading to decreased absorption intensity.

Hyperchromic effect. An effect leading to increased absorption intensity.

λ_{max}. The wavelength of an absorption maximum.

ε. The extinction coefficient defined by Eq. 1.2.

$E_{1\,cm}^{1\,\%}$. Absorption [$\log_{10}(I_0/I)$] of a 1 per cent solution in a cell with a 1 cm path length. This is used in place of ε when the molecular weight of a compound is not known, or when a mixture is being examined.

Isosbestic point. A point common to all curves produced in the spectra of a compound taken at several pH values.

1.13   Conjugated dienes

The energy levels of butadiene have been illustrated in Fig. 1.1. The transition *y* gives rise to strong absorption at 217 nm (ε 21 000). Alkyl substitution extends the chromophore, in the sense that there is a small interaction between the σ-bonded electrons of the alkyl group and the π-bond system. The result is a small red shift with alkyl substitution, just as there is a red shift (though a relatively large one) in going from an isolated double bond to a conjugated diene.

Fortunately the effect of alkyl substitution, in dienes at least, is additive; and a few rules suffice to predict the position of absorption in open chain dienes and dienes in six-membered rings. Open chain dienes exist normally in the s-*trans* conformation, while homoannular dienes must be in the s-*cis* conformation. These conformations are illustrated in the part structures **1** (heteroannular diene) and **2** (homoannular diene). It is not entirely clear why, but the s-*cis* conformation leads to longer wavelength absorption than does the s-*trans* conformation. Also, due to the shorter distance between the ends of the chromophore, s-*cis* dienes give maxima of lower intensity ($\varepsilon \sim 10\,000$) than the maxima of s-*trans* dienes ($\varepsilon \sim 20\,000$).

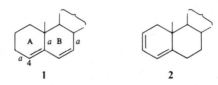

1   2

The actual rules for predicting the absorption of open chain and six-membered ring dienes were first made by Woodward in 1941. Since that time they have been modified by Fieser and by Scott as a result of experience with a very large number of dienes and trienes. The modified rules are given in Table 1.3.

For example the diene (**1**) would be calculated to have a maximum at 234 nm by the following addition:

Parent value		214 nm
Three-ring residues (marked a)	$3 \times 5 =$	15 nm
One exocyclic double bond (the Δ^4 bond is exocyclic to ring B)		5 nm
Total		234 nm

An observed value is 235 nm ($\varepsilon = 19\,000$).

Table 1.3   Rules for diene and triene absorption

Value assigned to parent heteroannular or open chain diene	214 nm
Value assigned to parent homoannular diene	253 nm
Increment for	
(a) each alkyl substituent or ring residue	5 nm
(b) the exocyclic nature of any double bond	5 nm
(c) a double bond extension	30 nm
(d) auxochrome—OAcyl	0 nm
—OAlkyl	6 nm
—SAlkyl	30 nm
—Cl, —Br	5 nm
—NAlkyl$_2$	60 nm
λ_{calc}	Total

(Reprinted with permission from A. I. Scott, *Interpretation of the Ultraviolet Spectra of Natural Products*, Pergamon Press, Oxford, 1964.)

By similar calculation, the diene (**2**) would be expected to have a maximum at 273 nm, and does actually have one at 275 nm. Though ethanol is the usual solvent, change of solvent has little effect. The actual appearance of the spectrum of a simple conjugated diene with the chromophore of **1** is illustrated in Fig. 1.2. The discussion above has referred to the highest intensity band and, indeed, the weaker bands are not always apparent.

There are a large number of exceptions to the rules, where special factors can operate. Distortion of the chromophore may lead to red or blue shifts, depending on the nature of the distortion.

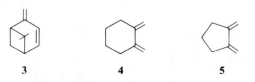

3   **4**   **5**

The strained molecule verbenene (**3**) has a maximum at 245.5 nm, whereas the usual calculation gives a value of 229 nm. The diene (**4**) might be expected to have a maximum at 273 nm; but distortion of the chromophore, presumably out of planarity with consequent loss of conjugation, causes the maximum to be as low as 220 nm with a similar loss in intensity (ε 5500). The diene (**5**), in which coplanarity of the diene is more likely, gives a maximum at 248 nm (ε 15 800) showing that this is so although it still does not agree with the expected value. Change of ring size in the case of simple homoannular dienes also leads to departures from the predicted value of 263 nm as

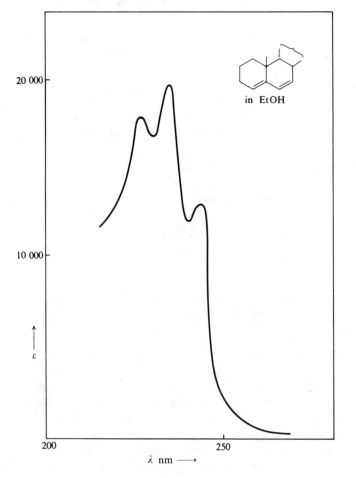

in EtOH

Fig. 1.2

follows: cyclopentadiene, 238.5 nm (ε 3400); cycloheptadiene, 248 nm (ε 7500); while cyclohexadiene is close at 256 nm (ε 8000). The lesson, an important one, is that when the ultraviolet spectrum of an unknown compound is to be compared with that of a model compound, then the choice of model must be a careful one. Allowance must be made for the likely shape of the molecule and for any unusual strain. Some general comments on the effect of steric hindrance to coplanarity are given in Sec. 1.26.

1.14   Polyenes

As the number of double bonds in conjugation increases, the wavelength of maximum absorption encroaches on the visible region. A number of subsidiary bands also appear and the intensity increases. Table 1.4 gives examples of the longest wavelength maxima of some simple conjugated polyenes, showing these trends.

Table 1.4   Longest wavelength maxima of some simple polyenes

	Me(CH=CH)$_n$Me		Ph(CH=CH)$_n$Ph	
n	λ_{max} nm	ε	λ_{max} nm	ε
3	274.5	30 000	358	75 000
4	310	76 500	384	86 000
5	342	122 000	403	94 000
6	380	146 500	420	113 000
7	401	–	435	135 000
8	411	–	–	–

The appearance of the spectra of some simple polyenes is illustrated in Fig. 1.3, which should be compared with the simpler spectrum of the diene in Fig. 1.2.

Several attempts, both empirical and theoretical, have been made to relate the principal or longest wavelength maximum with chain length. Some of the theoretical treatments have been based on the classical 'electron in the box' wave equation, in which the walls of the box are usually considered to be one average bond length beyond each end of the chromophore. This leads to the correct conclusion that increasing values of λ_{max} are found for increasing length in a conjugated polyene; quantitative predictions are, however, less satisfactory. The simple theory might indicate that as the chain length increases, the value of λ_{max} for long chains would increase proportionately, whereas in practice there is a convergence, which can be seen in Table 1.4. More sophisticated treatments, allowing for the variation in bond lengths between the double and single bonds, have been made and are described in Murrell's book. An interesting simplification is provided by the cyanine dye analogues (6) in which overlap leads to uniform bond lengths and bond orders along the polyene chain.

$$Me_2\overset{+}{N}=CH-(CH=CH)_n-NMe_2 \longleftrightarrow Me_2N-(CH=CH)_n-CH=\overset{+}{N}Me_2$$

6

Calculations based on the 'electron in the box' model lead to values very close to those observed: λ_{max} 309 ($n = 1$), 409 ($n = 2$) and 511 ($n = 3$) nm.

In a long-chain polyene, change from *trans* to *cis* configuration at one or more double bonds lowers both the wavelength and the intensity of the absorption maximum.

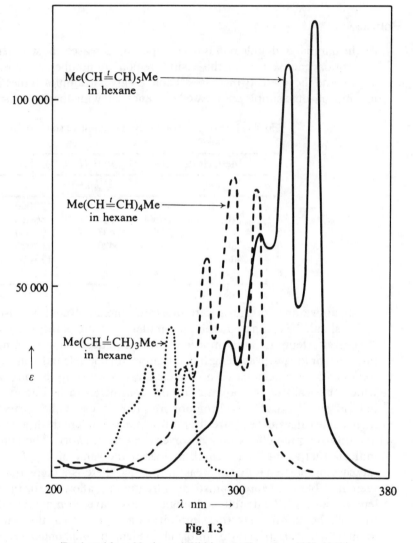

Fig. 1.3

(Replotted from Nayler and Whiting, *J. Chem. Soc.*, 1955, 3042.)

1.15   Polyeneynes and poly-ynes

As a result of interest in natural polyeneynes and poly-ynes, the ultraviolet spectra of many such compounds are known and have been of considerable use in the elucidation of structure. The characteristic spiky appearance of the spectra has been very helpful during the screening of crude plant extracts for acetylenic compounds. When more than two triple bonds are conjugated, the spectrum shows a characteristic series of low intensity bands ($\varepsilon \sim 200$) at intervals of 2300 cm^{-1} (note the *frequency* units, frequency being directly proportional to energy whereas wavelength is not) and high intensity bands ($\varepsilon \sim 10^5$) at intervals of 2600 cm^{-1}. The principal maxima in each group are shown in Table 1.5.

Table 1.5   Principal maxima of conjugated poly-ynes Me(C≡C)$_n$Me

n	λ_{max}, nm	ε	λ_{max}, nm	ε
2	–	–	250	160
3	207	135 000	306	120
4	234	281 000	354	105
5	260.5	352 000	394	120
6	284	445 000	–	–

Figure 1.4 shows the spectrum which is, like a fingerprint, diagnostic of the triyne-ene chromophore present in the dehydromatricaria ester (**7**).

$$Me—C≡C—C≡C—C≡C—CH\overset{t}{=}CH—CO_2Me$$

7

A similar compound (**8**), which has, effectively, an alkyl group in place of the carbomethoxyl group of the ester (**7**), shows a similar pattern shifted to the blue by about 15 nm, as shown on Fig. 1.4.

$$Me—C≡C—C≡C—C≡C—CH\overset{t}{=}CH—CH_2CH_2COEt$$

8

This is an example of the way in which an organic chemist deals with the comparison of ultraviolet spectra: most of the chromophore of **7** is present in **8**, and the latter will therefore continue to show the characteristic features of the former, with a small blue shift due to the relatively small loss of conjugation.

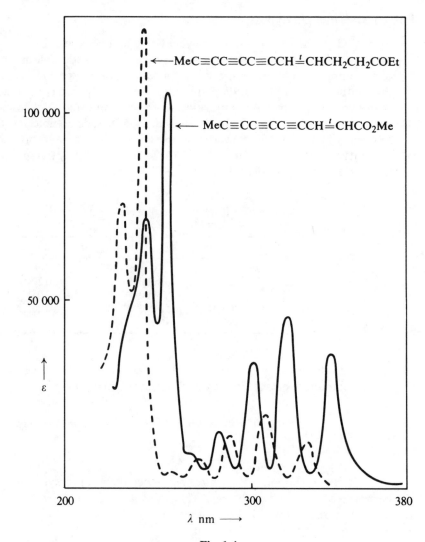

Fig. 1.4

(Replotted from Sörensen, Bruun, Holme and Sörensen, *Acta Chem. Scand.*, 1954, **8**, 28 and Bohlmann, Mannhardt and Viehe, *Chem. Ber.*, 1955, **88**, 365.)

1.16   Ketones and aldehydes; $\pi \rightarrow \pi^*$ transitions

Like the dienes considered in Sec. 1.13, $\alpha\beta$-unsaturated ketones and aldehydes have been the subject of much study and their absorption, too, is susceptible to prediction by a set of rules first formulated by Woodward and modified by Fieser and by Scott. The modified rules for calculating the expected position of the absorption maximum are given in Table 1.6.

Table 1.6 Rules for $\alpha\beta$-unsaturated ketone and aldehyde absorption

$$\overset{\delta}{C}=\overset{\gamma}{C}-\overset{\beta}{C}=\overset{\alpha}{C}-C=O$$ ε values are usually above 10 000 and increase with the length of the conjugated system.

Value assigned to parent $\alpha\beta$-unsaturated six-ring or acyclic ketone		215 nm
Value assigned to parent $\alpha\beta$-unsaturated five-ring ketone		202 nm
Value assigned to parent $\alpha\beta$-unsaturated aldehyde		207 nm
Increments for		
(a) a double bond extending the conjugation		30 nm
(b) each alkyl group or ring residue α		10 nm
β		12 nm
γ and higher		18 nm
(c) auxochromes		
(i) —OH α		35 nm
β		30 nm
δ		50 nm
(ii) —OAc α, β, δ		6 nm
(iii) —OMe α		35 nm
β		30 nm
γ		17 nm
δ		31 nm
(iv) —SAlk β		85 nm
(v) —Cl α		15 nm
β		12 nm
(vi) —Br α		25 nm
β		30 nm
(vii) —NR$_2$ β		95 nm
(d) the exocyclic nature of any double bond		5 nm
(e) homodiene component		39 nm
λ_{calc}^{EtOH}	Total	

For λ_{max}^{calc} in other solvents a solvent correction (Table 1.7) must be subtracted from the above value.

(Reprinted with permission from A. I. Scott, *Interpretation of the Ultraviolet Spectra of Natural Products*, Pergamon Press, Oxford, 1964.)

In this case, spectra are affected significantly by the solvent as a result of the change in polarity on excitation. A solvent correction (from Table 1.7) is subtracted from the calculated value (Table 1.6) to obtain the value expected for a solvent other than the standard solvent ethanol.

Table 1.7 Solvent corrections for $\alpha\beta$-unsaturated ketones

Solvent	Correction, nm
Ethanol	0
Methanol	0
Dioxan	+5
Chloroform	+1
Ether	+7
Water	−8
Hexane	+11
Cyclohexane	+11

(Reprinted with permission from A. I. Scott, *Interpretation of the Ultraviolet Spectra of Natural Products*, Pergamon Press, Oxford, 1964.)

For example, mesityl oxide (Me_2C=CHCOMe) may be calculated to have λ_{max} at $215 + (2 \times 12) = 239$ nm. The observed value is 237 nm (ε 12 600). A more complicated example, the trienone chromophore of **9**, would be calculated to have a maximum at 349 nm by the following addition.

9

Parent value	215 nm
β-substituent (marked a)	12 nm
ω-substituent (marked b)	18 nm
2 × extended conjugation	60 nm
Homoannular diene component (a special addition for this component when it is a linear part of the chromophore)	39 nm
Exocyclic double bond (the $\alpha\beta$-double bond is exocyclic to ring A)	5 nm
Total	349 nm

The observed values of λ_{max} are 230 nm (ε 18 000), 278 nm (ε 3720) and 348 nm (ε 11 000). As was the case with simple polyenes, the long chromophore present in this example gives rise to several peaks, with the longest wavelength peak in excellent agreement with prediction.

An important general principle is illustrated by the calculation for the cross-conjugated trieneone (**10**). In this case the main chromophore is the linear dieneone portion, since the Δ^5-double bond is not in the longest conjugated system. The calculation, along the lines above, gives a value of 324 nm. The observed values are 256 nm and 327 nm. The former might be due to the Δ^5-7-one system (λ_{calc} 244 nm), but a positive identification of this sort in a complicated system is largely unjustified.

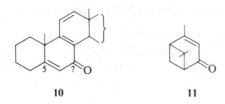

10   11

Certain special changes in structure, as noted in the case of dienes in Sec. 1.13, also lead to departures from the rules given above. The effect of the five-membered ring in cyclopentenones is accommodated in the rules; but when the carbonyl group is in a five-membered ring and the double bond is exocyclic to the five-membered ring, a parent value of about 215 nm holds. Another special case, verbenone (11), would be calculated to have a maximum at 239 nm but actually has a maximum at 253 nm, an increment for strain of 14 nm, close to the increment for the corresponding diene (3).

1.17   Ketones and aldehydes; n → π* transitions

Saturated ketones and aldehydes show a weak symmetry-forbidden band, in the 275–295 nm range ($\varepsilon \sim 20$), due to excitation of an oxygen lone-pair electron to the antibonding π-orbital of the carbonyl group. Aldehydes and the more heavily substituted ketones absorb at the upper end of this range. Polar substituents on the α-carbon atoms raise (when axial) or lower (when equatorial) the extremes of this range. When the carbonyl group is substituted by an auxochrome—as in an ester, an acid, or an amide—the π* orbital is raised but the n level of the lone-pair left largely unaltered. The result is that the n → π* transition of these compounds is shifted to the relatively inaccessible 200–215 nm range. The presence, therefore, of a weak band in the 275–295 nm region is positive identification of a ketone or aldehyde carbonyl group (nitro groups show a similar band and, of course, impurities must be absent). The low intensity of this transition is responsible for the ease with which the Cotton effect may be measured in studies of the optical rotatory dispersion of ketones.

αβ-unsaturated ketones show a slightly stronger n → π* band or series of bands ($\varepsilon \sim 100$) in the 300–350 nm range. The precise position of these bands is not predictable from the extent of alkylation, but is a regular function of the conformation of γ-substituents, axially substituted isomers absorbing at longer wavelengths than equatorially substituted isomers.

The position and intensity of n → π* bands are also influenced by transannular interactions (see Sec. 1.25) and by solvent (see Sec. 1.9).

The n → π* transitions of α-diketones in the diketo form give rise to two bands, one in the usual region near 290 nm ($\varepsilon \sim 30$) and a second (ε 10–30), which stretches into the visible in the 340–440 nm region and gives rise to the yellow colour of some of these compounds. (See also quinones in Sec. 1.23, quinones being α-, or vinylogous α-, diketones.)

1.18   αβ-unsaturated acids, esters, nitriles and amides

αβ-unsaturated acids and esters follow a trend similar to that of the ketones but at slightly shorter wavelength. The rules for alkyl substitution, summarized by Nielsen, are given in Table 1.8. The change in going from acid to ester is usually not more than 2 nm.

Table 1.8   Rules for αβ-unsaturated acids' and esters' absorption (ε values are usually above 10 000)

β-monosubstituted	208 nm
αβ- or ββ-disubstituted	217 nm
αββ-trisubstituted	225 nm
Increment for	
(a) a double bond extending the conjugation	30 nm
(b) the exocyclic nature of any double bond	5 nm
(c) when the double bond is endocyclic in a five- or seven-membered ring	5 nm
λ_{calc} Total	

αβ-unsaturated nitriles have been little studied but usually come slightly below the corresponding acids.

αβ-unsaturated amides have maxima lower than the corresponding acids, usually near 200 nm ($\varepsilon \sim 8000$).

αβ-unsaturated lactams have an additional band at 240–250 nm ($\varepsilon \sim 1000$).

1.19   The benzene ring

Benzene absorbs at 184 (ε 60 000), 203.5 (ε 7400) and 254 (ε 204) nm in hexane solution; it is illustrated by the solid line in Fig. 1.5. The latter band, sometimes called the *B*-band, shows vibrational fine structure. Although a 'forbidden' band, it owes its appearance to the loss of symmetry caused by molecular vibrations; indeed, the $0 \rightarrow 0$ transition (the transition between the ground state vibrational energy level of the electronic ground state to the ground state vibrational energy level of the electronic excited state) is not observed.

When the aromatic ring is substituted by alkyl groups, for example, or is an aza analogue such as pyridine, the symmetry is lowered; the $0 \rightarrow 0$ transition is then observed, although the spectrum is little changed otherwise. The presence of fine structure resembling that shown in Fig. 1.5 is characteristic of the simpler aromatic molecules.

When, however, the benzene ring is substituted by lone-pair donating or by π-bonded systems, the chromophore is extended more usefully; unfortunately, quantitative prediction of the effects of various substituents is not always possible in the manner so successful with dienes and unsaturated ketones. Section 1.20 gives an account of some of the trends observed in compounds containing a substituted benzene ring.

1.20   Substituted benzene rings

Table 1.9, giving the wavelength of absorption maxima in the spectra of a range of monosubstituted benzenes, shows how, as usual, the wavelength and intensity of the absorption peaks increase with an increase in the extent of the chromophore. As more and more conjugation is added to the benzene ring, the band originally at 203.5 nm (sometimes called the K-band) effectively 'moves' to longer wavelength, and moves 'faster' than the B-band, which was originally at 254 nm, eventually overtaking it. This can be seen in the two other spectra recorded on Fig. 1.5: benzoic acid (the dashed line) shows the K-band at 230 nm with the B-band still clearly visible at 273 nm; but with the longer chromophore of cinnamic acid (dotted line) the K-band has moved to 273 nm and the B-band is completely submerged. In the latter case, we can see how the even stronger band, originally at 184 nm, has also moved, but has still not reached the accessible region. It is responsible for what is called end absorption; that is, the long-wavelength side of an absorption peak, the maximum of which is below the range of the instrument.

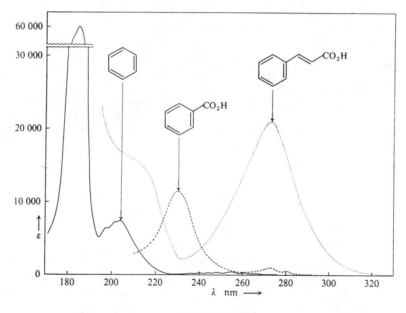

Fig. 1.5

Table 1.9   Absorption maxima of the substituted benzene rings Ph—R

R	λ_{max}, nm (ε) (solvent H_2O or MeOH)					
—H	203.5	(7 400)	254	(204)		
—NH$_3^+$	203	(7 500)	254	(160)		
—Me	206.5	(7 000)	261	(225)		
—I	207	(7 000)	257	(700)		
—Cl	209.5	(7 400)	263.5	(190)		
—Br	210	(7 900)	261	(192)		
—OH	210.5	(6 200)	270	(1 450)		
—OMe	217	(6 400)	269	(1 480)		
—SO$_2$NH$_2$	217.5	(9 700)	264.5	(740)		
—CN	224	(13 000)	271	(1 000)		
—CO$_2^-$	224	(8 700)	268	(560)		
—CO$_2$H	230	(11 600)	273	(970)		
—NH$_2$	230	(8 600)	280	(1 430)		
—O$^-$	235	(9 400)	287	(2 600)		
—NHAc	238	(10 500)				
—COMe	245.5	(9 800)				
—CH=CH$_2$	248	(14 000)	282	(750)	291	(500)
—CHO	249.5	(11 400)				
—Ph	251.5	(18 300)				
—OPh	255	(11 000)	272	(2 000)	278	(1 800)
—NO$_2$	268.5	(7 800)				
—CH$\overset{t}{=}$CHCO$_2$H	273	(21 000)				
—CH$\overset{t}{=}$CHPh	295.5	(29 000)				

(Most values taken with permission from H. H. Jaffé and M. Orchin, *Theory and Applications of Ultraviolet Spectroscopy*, Wiley, New York, 1962.)

In disubstituted benzenes, two situations are important. When electronically complementary groups, such as amino and nitro, are situated *para* to each other as in **12**, there is a pronounced red shift in the main absorption band, compared to the effect of either substituent separately, due to the extension of the chromophore from the electron donating group to the electron withdrawing group through the benzene ring (**12**, arrows). Alternatively, when two groups are situated *ortho* or *meta* to each other or when the *para* disposed groups are not complementary, as in **13**, then the observed spectrum is usually closer to that of the separate, non-interacting, chromophores.

These principles are illustrated by the examples in Table 1.10. The values in this table should be compared with each other and with the values for the single substituents separately given in Table 1.9.

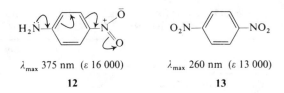

λ_{max} 375 nm (ε 16 000) λ_{max} 260 nm (ε 13 000)

12   **13**

In particular it should be noted that those compounds with non-complementary substituents, or with an *ortho* or *meta* substitution pattern, actually have a band (though a much weaker one) at longer wavelength than the compounds with interacting *para* disubstituted substituents. This fact is not in accord with the simple resonance picture; neither is the similarity of the *ortho* to the *meta* disubstituted cases. This is another case in which the molecular orbital theory (too complicated to be introduced here but dealt with in Murrell's book) gives a better picture.

Table 1.10   Absorption maxima of the disubstituted benzene rings $R-C_6H_4-R'$

R	R'	Orientation	λ_{max}^{EtOH}, nm (ε)		
—OH	—OH	*o*	214 (6 000)	278 (2 630)	
—OMe	—CHO	*o*	253 (11 000)	319 (4 000)	
—NH$_2$	—NO$_2$	*o*	229 (16 000)	275 (5 000)	405 (6 000)
—OH	—OH	*m*	277 (2 200)		
—OMe	—CHO	*m*	252 (8 300)	314 (2 800)	
—NH$_2$	—NO$_2$	*m*	235 (16 000)	373 (1 500)	
—Ph	—Ph	*m*	251 (44 000)		
—OH	—OH	*p*	225 (5 100)	293 (2 700)	
—OMe	—CHO	*p*	277 (14 800)		
—NH$_2$	—NO$_2$	*p*	229 (5 000)	375 (16 000)	
—Ph	—Ph	*p*	280 (25 000)		

In the case of disubstituted benzene rings in which the electron donating group is complemented by an electron withdrawing carbonyl group, some quantitative assessments may be made. These apply to the compounds $R-C_6H_4-COX$ in which X is alkyl, H, OH, or OAlkyl, and refer to the strongest band in the accessible region; this is often the only measured band in the highly conjugated *para* disubstituted systems. The calculation is based on a parent value with increments for each substituent. Polysubstituted benzene rings should be treated with caution, particularly when the substitution might lead to steric hindrance preventing coplanarity of the carbonyl group and the ring. Table 1.11 gives the rules for this calculation. In the absence of steric hindrance to coplanarity, the calculated values are usually within 5 nm of the observed values.

Table 1.11 Rules for the principal band of substituted benzene derivatives R—C₆H₄—COX

	Orientation	λ_{calc}^{EtOH}, nm
Parent chromophore:		
X = alkyl or ring residue		246
X = H		250
X = OH or OAlkyl		230
Increment for each substituent:		
R = alkyl or ring residue	o, m	3
	p	10
R = OH, OMe, OAlkyl	o, m	7
	p	25
R = O⁻	o	11
	m	20
	p	78
R = Cl	o, m	0
	p	10
R = Br	o, m	2
	p	15
R = NH₂	o, m	13
	p	58
R = NHAc	o, m	20
	p	45
R = NHMe	p	73
R = NMe₂	o, m	20
	p	85

(Reprinted with permission from A. I. Scott, *Interpretation of the Ultraviolet Spectra of Natural Products*, Pergamon Press, Oxford, 1964.)

A single example, that of 6-methoxytetralone (**14**), will show the method.

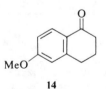

14

Parent value	246 nm
Ortho alkyl	3 nm
Para methoxyl	25 nm
λ_{calc}	274 nm

The maximum actually occurs at 276 nm (ε 16 500).

Other electron withdrawing groups, e.g. in nitriles and nitro compounds, show similar trends but with different and less well documented substituent effects.

1.21   Polycyclic aromatic hydrocarbons

The range of polycyclic aromatic hydrocarbons is too great for detailed consideration in this book. Their spectra are usually complicated, and for that reason are useful as fingerprints. This is particularly so in that the relatively non-polar substituents, such as alkyl and acetoxyl groups, have only a small effect on the shape and position of the absorption peaks of the parent hydrocarbon. The degradation products of natural materials often contain polycyclic nuclei which can be identified in this way as, for example, a phenanthrene or a perylene. The spectra of a typical series, naphthalene, anthracene and naphthacene, are illustrated in Fig. 1.6; the logarithmic ordinate should be noted.

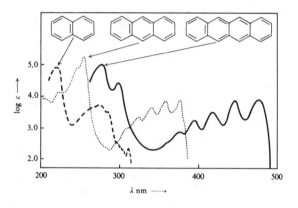

Fig. 1.6

(Reprinted with permission from R. A. Friedel and M. Orchin, *Ultraviolet Spectra of Aromatic Compounds*, Wiley, New York, 1951.)

Fortunately, the collections of spectra mentioned in Sec. 1.11 show the actual spectra of a great many of the known aromatic systems and make the identification of such systems a relatively simple matter.

1.22   Heteroaromatic compounds

The range of heteroaromatic compounds is too great for detailed consideration in this book. In general they resemble the spectra of their corresponding hydrocarbons, but only in the crudest way. The heteroatom, whether like that in a pyrrole or that in a pyridine, leads to pronounced substituent effects which depend on the electron donating or withdrawing effect of the substituent and the heteroatom and on their orientation. The effects of these factors are predictable, in a qualitative way, using the same sorts of criteria as were used in Sec. 1.20 when considering the effects of more than one substituent on a benzene ring. For example, a simple pyrrole (15) and a pyrrole with an electron withdrawing substituent (16) have strikingly different absorption maxima. The conjugation present from the nitrogen lone pair through the pyrrole ring to the carbonyl group increases the length of the chromophore and leads to longer wavelength absorption. The following illustrations of heterocyclic systems give some indication of the spectra observed.

15

λ_{max}^{EtOH} 203 nm
(ε 5670)

16

λ_{max}^{EtOH} 262 nm
(ε 12 000)

17

λ_{max}^{EtOH} 245 nm
(ε 4800)

18

λ_{max}^{EtOH} 300 nm
(ε 5000)

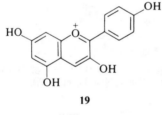

19

λ_{max}^{MeOH} 520 nm

20

$\lambda_{max}^{CHCl_3}$ 245 nm (ε 12 000)
275 nm (ε 2800)
282 nm (ε 3020)

21

$\lambda_{max}^{cyclohexane}$ 220 nm (ε 26 000)
262 nm (ε 6310)
280 nm (ε 5620)
288 nm (ε 4170)

22

$\lambda_{max}^{CHCl_3}$ 218 nm (ε 79 000)
266 nm (ε 3900)
305 nm (ε 2000)
318 nm (ε 3000)
Compare these values with the spectrum of naphthalene in Fig. 1.6.

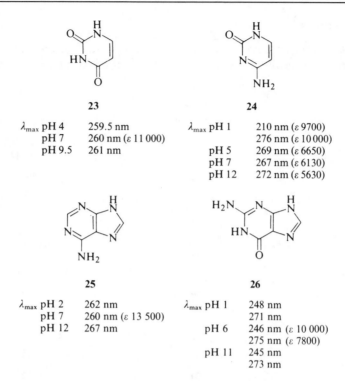

23

λ_{max} pH 4   259.5 nm
     pH 7   260 nm (ε 11 000)
     pH 9.5   261 nm

24

λ_{max} pH 1   210 nm (ε 9700)
            276 nm (ε 10 000)
     pH 5   269 nm (ε 6650)
     pH 7   267 nm (ε 6130)
     pH 12   272 nm (ε 5630)

25

λ_{max} pH 2   262 nm
     pH 7   260 nm (ε 13 500)
     pH 12   267 nm

26

λ_{max} pH 1   248 nm
            271 nm
     pH 6   246 nm (ε 10 000)
            275 nm (ε 7800)
     pH 11   245 nm
            273 nm

In the case of potentially tautomeric molecules the change in the absorption maxima with the change of pH is due sometimes to a change in the chromophore as a result of the tautomerism and sometimes to simple protonation or deprotonation. This point is mentioned here in order to stress the importance of careful control of the medium in which spectra are taken. The changes in absorption maxima with change of pH are very useful diagnostically since they serve in some systems to identify the pattern of substitution. The stable tautomeric species have been identified, using ultraviolet spectroscopy. For example, the 2-hydroxypyridine (**27**, R = H):pyrid-2-one (**28**, R = H) equilibrium has been shown to lie far to the right; the ultraviolet spectrum of the solution resembles that of a solution of *N*-methylpyrid-2-one (**28**, R = Me) and is different from that of 2-methoxypyridine (**27**, R = Me).

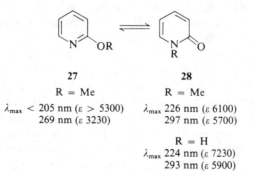

27

R = Me

λ_{max} < 205 nm (ε > 5300)
      269 nm (ε 3230)

28

R = Me

λ_{max} 226 nm (ε 6100)
     297 nm (ε 5700)

R = H

λ_{max} 224 nm (ε 7230)
     293 nm (ε 5900)

1.23   Quinones

A few representative quinones are illustrated below. The colour of the simpler members is due to the weak n → π^* transition, similar to that of α-diketones.

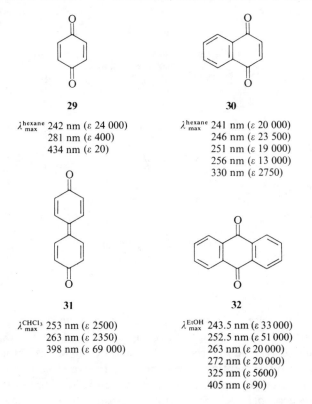

29

λ_{max}^{hexane} 242 nm (ε 24 000)
281 nm (ε 400)
434 nm (ε 20)

30

λ_{max}^{hexane} 241 nm (ε 20 000)
246 nm (ε 23 500)
251 nm (ε 19 000)
256 nm (ε 13 000)
330 nm (ε 2750)

31

$\lambda_{max}^{CHCl_3}$ 253 nm (ε 2500)
263 nm (ε 2350)
398 nm (ε 69 000)

32

λ_{max}^{EtOH} 243.5 nm (ε 33 000)
252.5 nm (ε 51 000)
263 nm (ε 20 000)
272 nm (ε 20 000)
325 nm (ε 5600)
405 nm (ε 90)

1.24   Porphyrins, chlorins and corroles

Our knowledge of the chemistry of these important groups of macrocyclic compounds has benefited considerably from the ease with which each class, in many of its various oxidation levels and with varying substitution patterns, can be recognized by the relative intensity of the four bands found in the visible region between 400 nm and 700 nm. In addition to these, a very strong sharp band (the Soret band) occurs near 400 nm (ε 100 000). It is interesting that another conjugated macrocyclic aromatic system, [18]-annulene, shows a similar intense band at 369 nm (ε 303 000).

These compounds are mentioned here to stress the importance and usefulness of ultraviolet and visible spectroscopy in the study of groups of compounds possessing a long, complicated chromophore. Although little can be accomplished in such systems from a theoretical point of view, the very large number of model systems available makes an empirical approach quite straightforward and very rewarding. These remarks apply to a large number of systems which, for one reason or another, have been studied, but which cannot be dealt with in this chapter.

1.25   Non-conjugated interacting chromophores

Non-conjugated systems usually have little effect on each other; diphenyl methane has a spectrum similar to that of toluene; the cross conjugation of the trieneone (10) was successfully ignored when calculating the expected absorption maximum; and even diphenyl ether is not very different from anisole. However, several special cases of non-conjugated interaction are known, two examples of which are given below. The unsaturated ketones (33) show the n → π* and π → π* transitions shifted in opposite directions when X becomes more electronegative. Presumably the π* orbital

is raised by transannular interaction with the $\overset{+}{N}Me_2$ group, but since the n electron

is closer to the $\overset{+}{N}Me_2$ group in the excited state than in the ground state, the n → π*

transition is of lower energy. The diene (34) has absorption in the accessible ultraviolet whereas the isolated ethylenic double bond has no maximum above 190 nm.

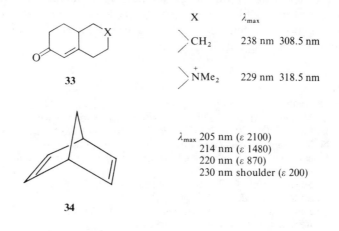

X	λ_{max}	
$>CH_2$	238 nm	308.5 nm
$>\overset{+}{N}Me_2$	229 nm	318.5 nm

33

λ_{max} 205 nm (ε 2100)
214 nm (ε 1480)
220 nm (ε 870)
230 nm shoulder (ε 200)

34

1.26   The effect of steric hindrance to coplanarity

Steric hindrance to coplanarity about a double bond, as in the hydrocarbon (35), raises the ground state energy level but leaves the excited state relatively unchanged (the latter is probably of lowest energy in the conformation in which the biphenyl systems are at right angles). The result [in this case a series of bands culminating at 458 nm (ε 23 000)] is a shift towards the red from what might have been expected.

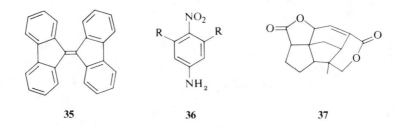

35   36   37

Mild steric hindrance to coplanarity about a single bond has only a small effect on the position and intensity of absorption maxima.

Medium steric hindrance to coplanarity about a single bond gives rise to a marked decrease in intensity but may also lead to either a blue shift or a red shift. For example, the absorption maximum of the nitroaniline **36** (R = Me) is at 385 nm (ε 4840), showing a red shift and marked reduction in intensity from that of the parent compound **36** (R = H) at 375 nm (ε 16 000). Another example, in the opposite direction, is that of 2,4,6-trimethylacetophenone absorbing at 242 nm (ε 3200), which is to be compared with the calculated value (Table 1.11) of 262 nm and with *p*-methylacetophenone which has a maximum at 252 nm (ε 15 000).

Extreme steric hindrance to coplanarity about a single bond leads to a situation with no overlap between the separated chromophores. The dilactone (**37**) produced from shellolic acid showed no maximum in the accessible ultraviolet region but on hydrolysis of the $\alpha\beta$-unsaturated lactone grouping an acid with λ_{max} 227 nm (ε 5500) was obtained. This shows that the steric constraint of the lactone ring prevents conjugation and that release of this constraint then allowed the overlap of the double bond and carbonyl orbitals.

Bibliography

TEXTBOOKS

S. F. Mason, Chapter 7, The Electronic Absorption Spectra of Heterocyclic Compounds, *Physical Methods in Heterocyclic Chemistry*, Vol. II, Academic Press, New York, 1963.

C. N. R. Rao, *Ultraviolet and Visible Spectroscopy*, Butterworths, London, 3rd Ed., 1975.

A. I. Scott, *Interpretation of the Ultraviolet Spectra of Natural Products*, Pergamon Press, Oxford, 1964.

E. S. Stern and C. J. Timmons, *Introduction to Electronic Absorption Spectroscopy*, Arnold, London, 3rd Ed., 1970.

W. West (Ed.), *Technique of Organic Chemistry*, Vol. IX, Chemical Applications of Spectroscopy, Interscience, New York, 2nd Ed., 1968.

THEORETICAL TREATMENTS

G. R. Barrow, *Introduction to Molecular Spectroscopy*, McGraw-Hill, New York, 1962.

E. F. H. Brittain, W. O. George and C. H. J. Wells, *Introduction to Molecular Spectroscopy*, Academic Press, London, 1970.

R. E. Dodd, *Chemical Spectroscopy*, Elsevier, Amsterdam, 1962.

H. H. Jaffé and M. Orchin, *Theory and Applications of Ultraviolet Spectroscopy*, Wiley, New York, 1962.

J. N. Murrell, *The Theory of the Electronic Spectra of Organic Molecules*, Methuen, London, 1963.

2. Infrared spectra

2.1 *Introduction.* 2.2 *Preparation of samples and examination in an infrared spectrometer.* 2.3 *Fourier transform infrared spectroscopy.* 2.4 *Examination in a Raman spectrometer.* 2.5 *Selection rules.* 2.6 *The infrared spectrum.* 2.7 *The use of the tables of characteristic group frequencies.* 2.8 *Correlation charts.* 2.9 *Absorption frequencies of single bonds to hydrogen.* 2.10 *Absorption frequencies of triple bonds and cumulated double bonds.* 2.11 *The aromatic overtone and combination region, 2000–1600 cm⁻¹.* 2.12 *Absorption frequencies of the double bond region.* 2.13 *Groups absorbing in the fingerprint region.* 2.14 *Examples of infrared spectra. Bibliography.*

2.1 Introduction

The energy of most molecular vibrations corresponds to that of the infrared region of the electromagnetic spectrum. Molecular vibrations may be detected and measured either in an infrared spectrum or indirectly in a Raman spectrum. The most useful vibrations, from the point of view of the organic chemist, occur in the narrower range of 2.5–16 μm (1 μm = 10^{-4} cm) which most infrared spectrometers cover. The position of an absorption band in the spectrum may be expressed in microns (μm), or very commonly—and throughout this book—in terms of the reciprocal of the wavelength, cm⁻¹. The usual range of an infrared spectrum is, therefore, between 4000 cm⁻¹ at the high frequency end and 625 cm⁻¹ at the low frequency end.

Functional groups have vibration frequencies, characteristic of that functional group, within well-defined regions of this range; these are summarized in Figs 2.3, 2.4, 2.5, and 2.6, and form the subject matter of this chapter. We will, however, postpone discussion of the regions in which functional groups absorb until we have established the very great ease with which samples are prepared and spectra taken. The fact that many functional groups can be identified by their characteristic vibration frequencies makes the infrared spectrum the simplest, most rapid, and often most reliable means for assigning a compound to its class.

2.2 Preparation of samples and examination in an infrared spectrometer

The spectrometer consists of a source of infrared light, emitting radiation throughout the whole frequency range of the instrument. This light is split into two beams of equal intensity, and one beam is arranged to pass through the sample to be examined. If the frequency of a vibration of the sample molecule falls within the range of the instrument, the molecule may absorb energy of this frequency from the light. The spectrum is, therefore, scanned by comparing the intensity of the two beams after one has passed through the sample to be examined. The wavelength range over which the comparison is made is spread out in the usual way with a prism or grating. The whole operation is done automatically in such a way that the usual finished spectrum consists of a chart showing downward peaks, corresponding to absorption, plotted against wavelength or frequency. To allow for variations in the spectrometer, spectra are often calibrated against accurately known bands of the spectrum of polystyrene, the peaks of one or more of these bands being superimposed on the spectrum which is to be taken (see Fig. 2.2).

Compounds may be examined in the vapour phase, as pure liquids, in solution, and in the solid state (see Fig. 2.13).

In the vapour phase.   The vapour is introduced into a special cell, usually about 10 cm long, which can then be placed directly in the path of one of the infrared beams. The end walls of the cell are usually made of sodium chloride, which is transparent to infrared. Most organic compounds have too low a vapour pressure for this phase to be useful.

As a liquid.   A drop of the liquid is squeezed between flat plates of sodium chloride (transparent throughout the $4000–625\ \mathrm{cm}^{-1}$ region). This is the simplest of all procedures.

In solution.   The compound is dissolved to give, typically, a 1–5 per cent solution in carbon tetrachloride or, for its better solvent properties, alcohol-free chloroform. This solution is introduced into a special cell, 0.1–1 mm thick, made of sodium chloride. A second cell of equal thickness, but containing pure solvent, is placed in the path of the other beam of the spectrometer in order that solvent absorptions should be balanced. Spectra taken in such dilute solutions in non-polar solvents are generally the most desirable, because they are normally better resolved (see Fig. 2.13c) than spectra taken on solids, and also because intermolecular forces, which are especially strong in the crystalline state, are minimized. On the other hand, many compounds are not soluble in non-polar solvents, and all solvents absorb in the infrared; when the solvent absorption exceeds about 65 per cent of the incident light, spectra cannot be taken because insufficient light is transmitted to work the detection mechanism efficiently. Carbon tetrachloride and chloroform, fortunately, absorb over 65 per cent of the incident light only in those regions (Fig. 2.1) which are of little interest in diagnosis. Other solvents, of course, may be used but the areas of usefulness in each case should be checked beforehand, taking account of the size of the cell being used. In rare cases aqueous solvents are useful; special calcium fluoride cells are used.

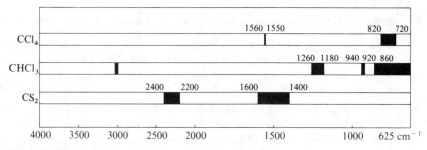

Fig. 2.1   Darkened areas are the regions in which the solvent cannot be used with a 0.2 mm cell.

In the solid state.   About 1 mg of a solid is finely ground in a small agate mortar with a drop of a liquid hydrocarbon (Nujol, Kaydol) or, if C—H vibrations are to be examined, with hexachlorobutadiene. The mull is then pressed between flat plates of sodium chloride. Alternatively, the solid is ground with 10–100 times its bulk of pure potassium bromide and the mixture pressed into a disc using a special mould and a hydraulic press. The use of KBr eliminates the problem (usually not troublesome) of bands due to the mulling agent (see Figs 2.13a and 2.13b) and tends, on the whole, to give rather better spectra, except that a band at 3450 cm^{-1}, from the OH group of traces of water, almost always appears (see Fig. 2.9). Due to intermolecular interactions, band positions in solid state spectra are often different from those of the corresponding solution spectra. This is particularly true of those functional groups which take part in hydrogen bonding. On the other hand, the number of resolved lines is often greater in solid state spectra (see Figs 2.13), so that comparison of the spectra of, for example, synthetic and natural samples in order to determine identity is best done in the solid state. This is only true, of course, when the same crystalline modification is in use; racemic, synthetic material and optically active, natural material, for example, should be compared in solution.

2.3   Fourier transform infrared spectroscopy (FTIR)

A new method of taking an infrared spectrum has come into use more recently. Light covering the whole frequency range, typically 5000–400 cm^{-1}, is split into two beams. Either one beam is passed through the sample, or both are passed, but one beam is made to traverse a longer path than the other. Recombination of the two beams produces an interference pattern that is the sum of all the interference patterns created by each wavelength in the beam. By systematically changing the difference in the two paths, the interference patterns change to produce a detected signal varying with optical path difference. This pattern is known as the *interferogram*, and looks nothing like a spectrum. However, Fourier transformation of the interferogram, using a computer built into the machine, converts it into a plot of absorption against wavenumber which resembles the usual spectrum obtained by the traditional method, described in Sec. 2.2. There are several advantages to FTIR over the traditional method, and few disadvantages. Because it is not necessary to scan each wavenumber successively, the whole spectrum is measured in at most a few seconds. Because it is not dependent upon a slit and a prism or grating, high resolution in FTIR is easier to obtain without sacrificing sensitivity. FTIR is especially useful for examining small samples (several scans can be added together) and for taking the spectrum of compounds produced in the outflow of a chromatograph (the conventional method, taking several minutes, requires that the sample be collected first). Finally, the digital form in which the data are handled in the computer allows the spectrum of a pure compound to be subtracted easily from that of a mixture to reveal the spectrum of the other component or components of the mixture. This list by no means exhausts the advantages of FTIR. For the purposes of this book, however, the way in which infrared spectra are taken does not affect the job of interpreting them. The spectra look and are very similar.

2.4   Examination in a Raman spectrometer

Raman spectra are generally taken on machines using laser sources, and the quantity of material needed is now of the order of a few mg. A liquid or a concentrated solution is irradiated with the monochromatic light, and the *scattered* light is examined through a spectrometer using photoelectric detection. Most of the scattered light consists of the parent line produced by absorption and re-emission. Much weaker lines, which constitute the Raman spectrum, occur at lower and higher energy and are due to absorption and re-emission of light coupled with vibrational excitation or decay respectively. The difference in frequency between the parent line and the Raman line is the frequency of the corresponding vibration.

Raman spectroscopy is not used by organic chemists for structure determination as routinely as is infrared spectroscopy, but for the detection of certain functional groups (see Sec. 2.4), and for the *analysis* of mixtures—of deuterated compounds for example—it has found much use.

2.5   Selection rules

Infrared light is absorbed when the oscillating dipole moment (due to a molecular vibration) interacts with the oscillating electric vector of the infrared beam. A simple rule for deciding if this interaction (and hence absorption of light) occurs is that the dipole moment at one extreme of a vibration must be different from the dipole moment at the other extreme of the vibration. In the Raman effect a corresponding interaction occurs between the light and the molecule's polarizability, resulting in different selection rules.

The most important consequence of these selection rules is that in a molecule with a centre of symmetry those vibrations symmetrical about the centre of symmetry are active in the Raman and inactive in the infrared; those vibrations which are not centrosymmetric are inactive in the Raman and usually active in the infared. This is doubly useful, for it means that the two types of spectrum are complementary; and the more easily obtained, the infrared, is the most informative for organic chemists, because most functional groups are not centrosymmetric.

The symmetry properties of a molecule in a solid can be different from those of an isolated molecule. This can lead to the appearance of infrared absorption bands in a solid state spectrum which would be forbidden in solution or in the vapour phase.

2.6   The infrared spectrum

A complex molecule has a large number of vibrational modes which involve the whole molecule. To a good approximation, however, some of these molecular vibrations are associated with the vibrations of individual bonds or functional groups (localized vibrations) while others must be considered as vibrations of the whole molecule.

The localized vibrations are either stretching, bending, rocking, twisting, or wagging. For example, the localized vibrations of the methylene group are

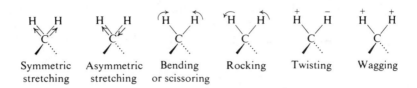

Symmetric stretching Asymmetric stretching Bending or scissoring Rocking Twisting Wagging

Many localized vibrations are very useful for the identification of functional groups.

The soggy vibrations of the molecule as a whole give rise to a series of absorption bands at low energy, below 1500 cm^{-1}, the positions of which are characteristic of that molecule. These bands make those localized vibrations which have frequencies below 1500 cm^{-1} less useful for diagnostic purposes since confusion of one with the other may occur. Frequently bands are observed which do not correspond to any of the fundamental vibrations of the molecule and are due to overtone bands and combination bands, the latter as a result of interaction between two or more vibrations. Occasionally these bands are useful diagnostically but more usually they supplement the region below 1500 cm^{-1}. The net result, when a spectrum has been taken, is a region above 1500 cm^{-1} showing absorption bands assignable to a number of functional groups and a region, characteristic of the compound in question and no other compound, containing many bands below 1500 cm^{-1}. This region, for obvious reasons, is called the fingerprint region. The use of the fingerprint region to confirm the identity of a compound with an authentic sample is considerably more reliable, in most cases, than the technique of taking a mixed melting point. Within the fingerprint region some bands assignable to functional groups do occur and may be used diagnostically; such identifications should be regarded as helpful rather than as definitive.

The regions in which functional groups absorb are summarized in Fig. 2.2, which also illustrates the very simple spectrum of the liquid paraffin Nujol, the mulling agent often used when taking the spectrum of a solid sample. Since Nujol possesses only C—H and C—C bonds, its spectrum shows features found in the majority of organic compounds.

The stretching vibrations of single bonds to hydrogen give rise to the absorption at the high frequency end of the spectrum as a result of the low mass of the hydrogen atom. Thereafter, the order of stretching frequencies follows the order: triple bonds at higher frequency than double bonds and double bonds higher than single bonds— on the whole the greater the strength of the bond between two similar atoms the higher the frequency of the vibration. Bending vibrations are of much lower frequency and usually appear in the fingerprint region below 1500 cm^{-1}. An exception is the N—H bending vibration which appears in the 1600 cm^{-1} region.

2.7 The use of the tables of characteristic group frequencies

Each of the three frequency ranges above 1500 cm^{-1} shown in Fig. 2.2. is expanded in the four charts Figs 2.3, 2.4, 2.5, and 2.6. These charts summarize the narrower ranges within which each of the functional groups absorbs. The absorption bands which are found in the fingerprint region and which are assignable to functional

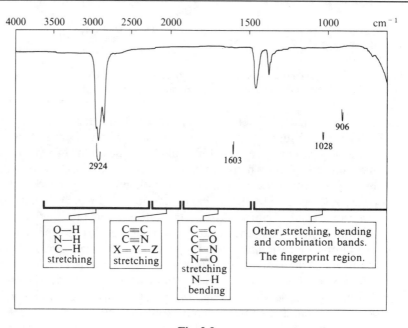

Fig. 2.2

groups are summarized in the chart Fig. 2.7; these latter correlations are occasionally useful, either because they are sometimes strong bands in otherwise featureless regions or because their absence may rule out incorrect structures. Following these summaries are Tables 2.1–2.23 arranged by functional groups roughly in order of their stretching frequencies. Where a functional group gives rise to absorption bands in addition to those due to stretching frequencies, their position is also mentioned in the table. This organization enables one to examine the main regions of the spectrum in turn, rather than to work backwards from a guess as to the functional groups present.

One could deal with the spectrum of an unknown as follows. Examine each of the three main regions of the spectrum covered by Figs 2.3–2.6; at this stage certain combinations of structures can be ruled out and some tentative conclusions reached. Where there is still ambiguity, the tables corresponding to those groups which might be present should be consulted, whereupon more detailed information should be available. It is well to be sure that the bands under consideration are of the appropriate intensity for the structure suspected. Some assistance in this task is provided at the end of this chapter, where nine infrared spectra are reproduced in order to show the usual appearance of a number of characteristically shaped bands.

Except where otherwise stated, band positions are given for dilute solution in non-polar solvents. Intensities in the infrared are less frequently recorded and less conveniently measured than is the case with ultraviolet spectra. Usually intensities are expressed subjectively as strong (s), medium (m), weak (w) and, in books such as this, variable (v). The position of all bands is given in cm^{-1}.

2.8 Correlation charts

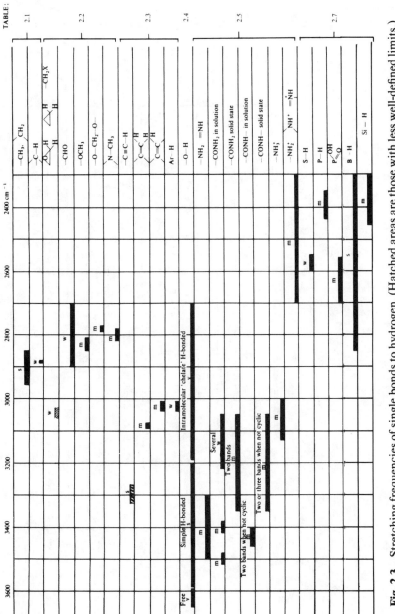

Fig. 2.3 Stretching frequencies of single bonds to hydrogen. (Hatched areas are those with less well-defined limits.)

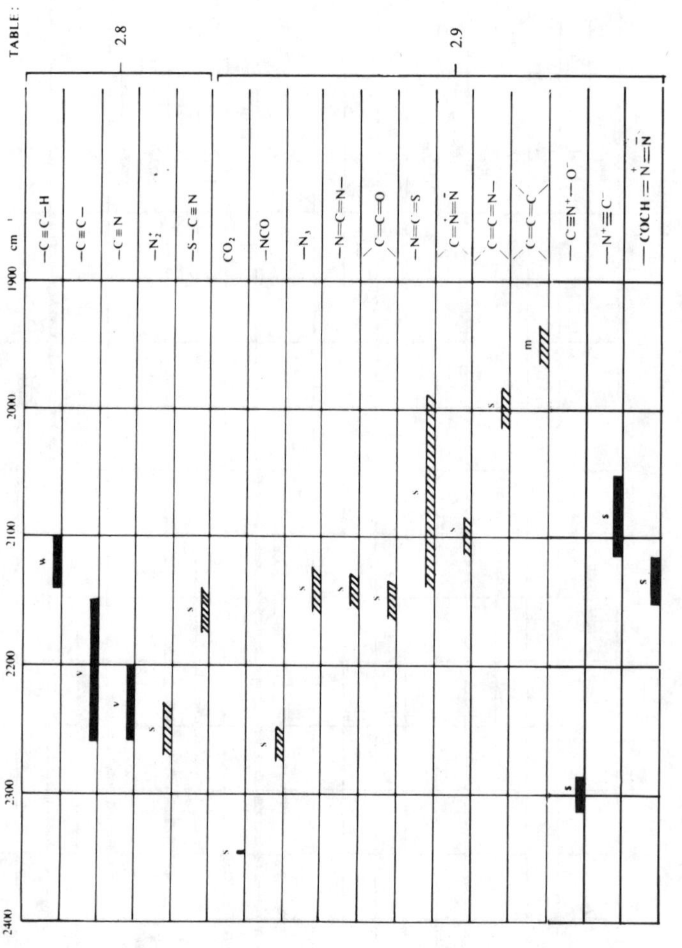

Fig. 2.4   Stretching frequencies of triple bonds and cumulated double bonds. (Hatched areas are those with less well-defined limits.)

Fig. 2.5 Double bond stretching and N—H bending frequencies. For carbonyl groups see Fig. 2.6. (Hatched areas are those with less well-defined limits.)

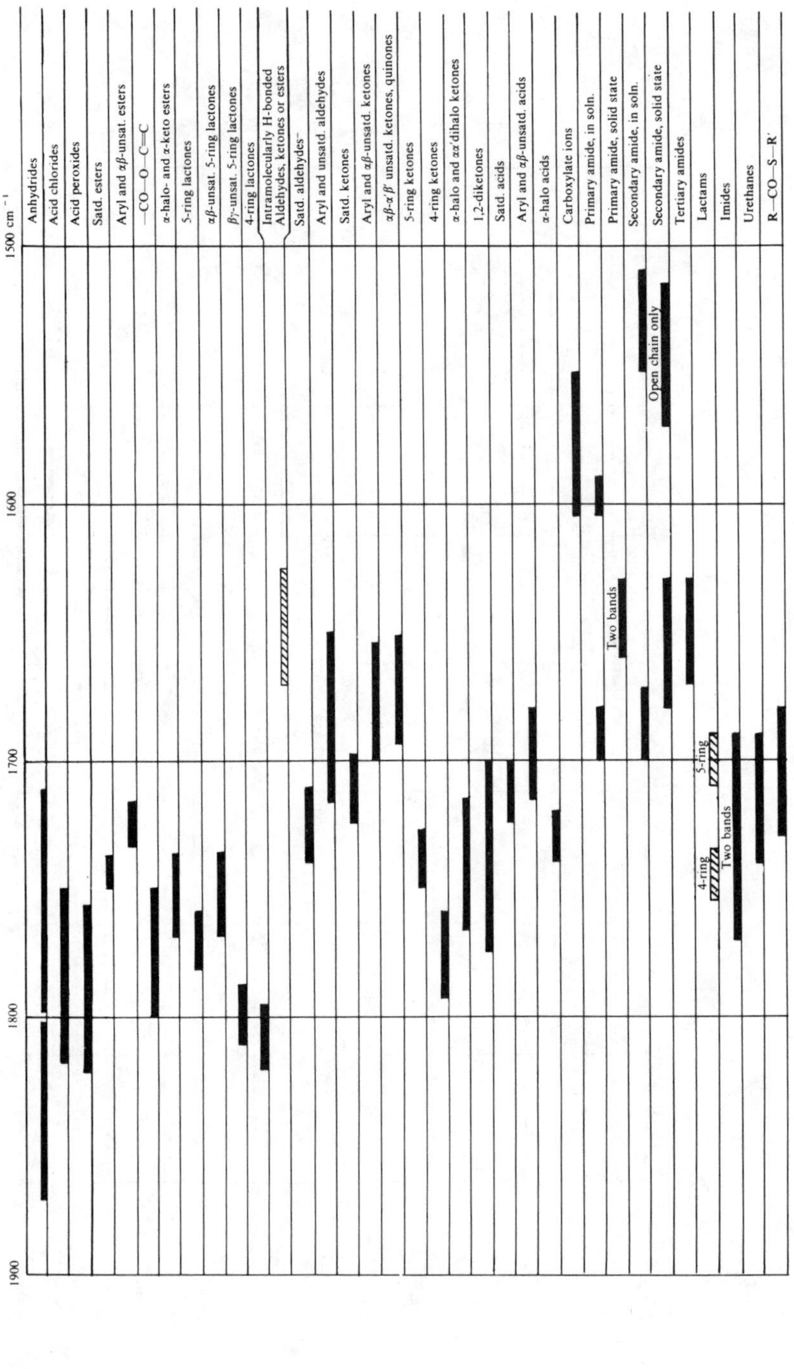

Fig. 2.6 Stretching frequencies of carbonyl groups. All values are found in Table 2.10. All bands are strong.

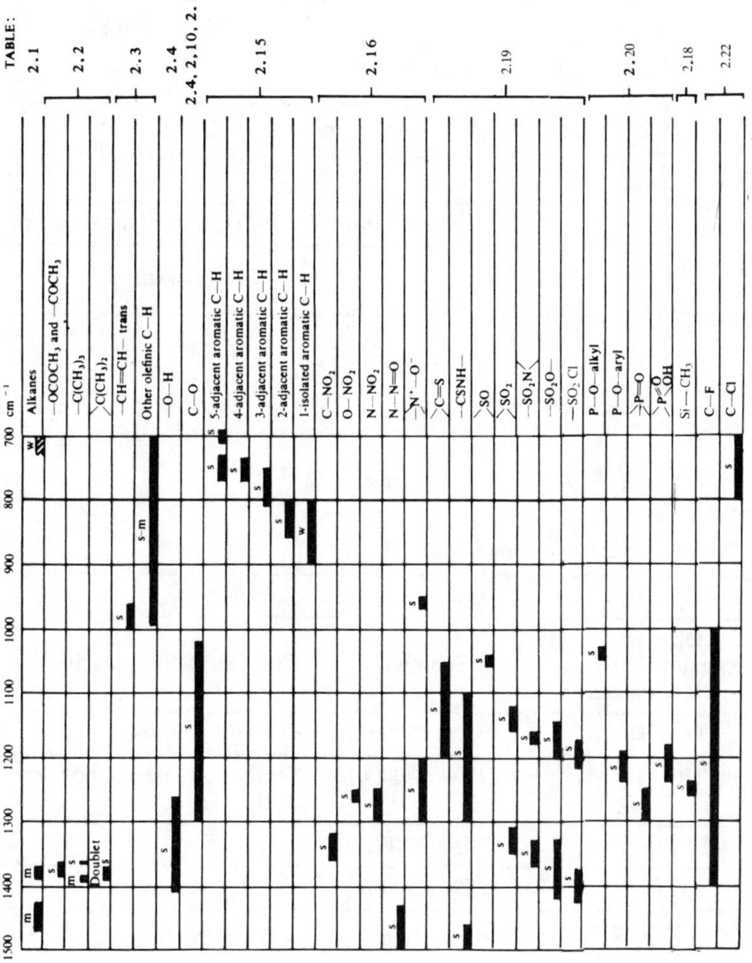

Fig. 2.7 Some characteristic bands in the fingerprint region.

2.9   Absorption frequencies of single bonds to hydrogen

Table 2.1   Saturated C—H and C—C

Group	Band	Remarks
$>$CH$_2$ —CH$_3$	2960–2850(s)	Two or three bands usually; $>$C—H stretching
$>$CH	2890–2880(w)	
$>$CH$_2$ —CH$_3$	1470–1430(m)	$>$C—H deformations
—CH$_3$	1390–1370(m)	—CH$_3$ symmetrical deformation
$>$CH$_2$	~720(w)	$>$CH$_2$ rocking

Table 2.2   Miscellaneous C—H

Group	Band	Remarks
Cyclopropane C—H Epoxide C—H —CH$_2$-halogen	~3050(w)	C—H stretching; cf. alkenes
—CO—CH$_3$	3100–2900(w)	Often very weak
—CHO	2900–2700(w)	Usually two bands, one near 2720 cm^{-1} (see Fig. 2.13a)
—O—CH$_3$	2850–2810(m)	
—O—CH$_2$—O—	2790–2770(m)	
N—CH$_2$ and N—CH$_2$—	2820–2780(m)	(See Fig. 2.12)
—C(CH$_3$)$_3$	1395–1385(m) 1365(s)	
$>$C(CH$_3$)$_2$	~1380(m)	A roughly symmetrical doublet (see Fig. 2.8)
—O—CO—CH —CO—CH$_3$	1385–1365(s) 1360–1355(s)	The high intensity of these bands often dominates this region of the spectrum

Table 2.3   Alkene and aromatic C—H. See also Table 2.13 and Table 2.14 for the corresponding double bond absorptions, and Table 2.15 for the aromatic C—H out-of-plane bending vibrations.

Group	Band	Remarks
—C≡C—H	~3300(s)	Sharp
$>C=C<$ with H, H	3095–3075(m)	C—H stretching; sometimes obscured by the much stronger bands of saturated C—H groups, which occur below 3000 cm^{-1}
$>C=C<$ with H	3040–3010(m)	(See Fig. 2.12)
Aryl—H	3040–3010(w)	Often obscured (but see Fig. 2.9)
R, H / C=C / H, R	970–960(s)	C—H out-of-plane deformation. When the double bond is conjugated with, for example, a C=O group this band is shifted towards 990 cm^{-1}
RCH=CH$_2$	995–985(s) and 940–900(s)	
R$_2$C=CH$_2$	895–885(s)	
R$_2$C=C< with H, R	840–790(m)	
H, H / C=C / R, R	730–675(m)	

Much is known about the precise position of the various CH, CH$_2$, and CH$_3$, symmetrical and unsymmetrical vibration frequencies. C—H bonds do not take part in hydrogen bonding and, therefore, their position is little affected by the state of measurement or their chemical environment. C—C vibrations, which absorb in the fingerprint region, are generally weak and not practically useful. Since most organic molecules posses alkane residues, the groups of saturated C—H absorption bands given above are of little diagnostic value; their general appearance may be seen in the Nujol spectrum (Fig. 2.2). The absence of saturated C—H absorption in a spectrum is, of course, diagnostic evidence for the absence of such a part structure in the corresponding compound (see Fig. 2.13a). Unsaturated and aromatic C—H stretching frequencies can be distinguished from the saturated C—H absorption since the latter occurs below 3000 cm^{-1} (see, however, Table 2.2) while the former gives rise to much less intense absorption above 3000 cm^{-1}. Alkene and aromatic C—H absorption is covered by Table 2.3.

A few special structural features in saturated C—H groupings give rise to characteristic absorption bands. These are summarized in Table 2.2 on page 40.

Table 2.4   Alcohol and phenol —O—H

Group	Band	Remarks
Water in solution	3710	
Free —OH	3650–3590(v)	Sharp; O—H stretching (see Fig. 2.8)
H-bonded —OH (solid, liquid, and dilute solution)	3600–3200(s)	Often broad but may be sharp for some intramolecular single bridge H bonds; the lower the frequency the stronger the H bond (see Fig. 2.8)
Intramolecular H-bonded —OH in chelate form (see also Table 2.10, carboxylic acids)	3200–2500(v)	Broad; the lower the frequency the stronger the H bond; sometimes so broad as to be overlooked
Water of crystallization (solid state spectra)	3600–3100(w)	Usually a weak band at 1640–1615 cm^{-1} also; water in trace amounts in KBr discs shows a broad band at 3450 cm^{-1} (see Fig. 2.9)
—O—H	1410–1260(s)	O—H bending
$\overset{|}{\underset{|}{C}}$—OH	1150–1040(s)	C—O stretching (see Fig. 2.8)

The value of the O—H stretching frequency has been used for many years as a test for and measure of the strength of hydrogen bonds. The stronger the hydrogen bond the longer the O—H bond, the lower the vibration frequency and the broader and the more intense the absorption band. The sharp free 'monomeric' band in the 3650–3590 cm^{-1} range can be observed in the vapour phase, in dilute solution or when such factors as steric hindrance prevent hydrogen bonding. Pure liquids, solids, and many solutions show only the broad, 'polymeric' band in the 3600–3200 range. Frequently liquid phase spectra show both bands (see Fig. 2.8).

Intramolecular hydrogen bonds of the non-chelate type, for example in 1,2-diols, show a sharp band in the range 3570–3450 cm^{-1}, the precise position being a measure of the strength of the hydrogen bond. A similar, though rather less sharp band is observed when the hydrogen bonding gives rise to dimers only. The 'polymeric' band is generally much broader. Distinctions can be made among the various possibilities by testing the effect of dilution; intramolecular hydrogen bonds are unaffected and the absorption band is, therefore, unaffected; while intermolecular hydrogen bonds are broken, leading to a decrease in the bonded O—H absorption and an increase in or the appearance of free O—H absorption. Spectra taken of samples in the solid state show only the broad strong band in the range 3400–3200 cm^{-1}.

The stretching frequencies of N—H bonds (see Table 2.5) can sometimes be confused with those of hydrogen bonded O—H frequencies. Due to their much weaker tendency to form hydrogen bonds, N—H absorption is usually sharper; moreover, N—H absorption is of weaker intensity and, in dilute solutions, it never gives rise to absorption as high as the free O—H range near 3600 cm^{-1}. Weak bands, overtones of the strong carbonyl absorption in the $1800-1600 \text{ cm}^{-1}$ region, also appear in the $3600-3200 \text{ cm}^{-1}$ region.

The effects of hydrogen bonding can be seen when a carbonyl group is the acceptor, for its characteristic stretching frequency is also lowered (see Table 2.10).

The characteristic series of bands in the $3000-2500 \text{ cm}^{-1}$ range produced by most carboxylic acids can be seen in Fig. 2.9, The highest frequency band is due to the O—H stretching vibration, and the other bands to combination vibrations. The bands are usually seen as a jagged series on the low frequency side of any C—H absorption which may be present. Combined with a carbonyl absorption in the correct region (Table 2.10) this series is very useful for the identification of carboxylic acids.

Table 2.5   Amine, imine, ammonium, and amide N—H stretching

Group	Band	Remarks
Amine and imine $>$N—H =N—H	3500–3300(m)	Primary amines show two bands in this range: the unsymmetrical and symmetrical stretching. Secondary amines absorb weakly. The pyrrole and indole N—H band is sharp (see Fig. 2.10)
—NH$_3$ Amino acids Amino salts	3130–3030(m) ~3000(m)	Values for solid state; broad; bands also (but not always) near 2500 and 2000 cm^{-1} (see text below Fig. 2.10)
$>$NH$_2$ $>$NH$^+$ =NH (with + above)	2700–2250(m)	Values for solid state; broad; due to the presence of overtone bands, etc.
Primary amide —CONH$_2$	~3500(m) ~3400(m)	Lowered ~150 cm^{-1} in the solid state and on H bonding; often several bands 3200–3050 cm^{-1} (see Fig. 2.11)
Secondary amide —CONH—	3460–3400(m) 3100–3070(w)	Two bands: lowered on H bonding and in the solid state (see Fig. 2.13a). Only one band with lactams A weak extra band with bonded and solid state samples (see Fig. 2.13a)

Much is known of amide N—H absorptions, the appearance of two bands being ascribed to forms **1** and **2**. The carbonyl region of many amides (Table 2.10) also shows two bands.

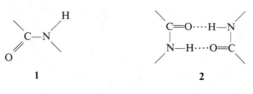

1   2

Hydrogen bonding lowers and broadens N—H stretching frequencies to a lesser extent than was the case with O—H groups. The intensity of N—H absorption is usually less than that of O—H absorption.

Table 2.6 N—H bending. See also Table 2.10 for amide absorptions in this region.

Group	Band	Remarks
—NH$_3$	1650–1560(m)	
>NH	1580–1490(w)	Often too weak to be noticed
—NH$_3^+$	1600(s) 1500(s)	Secondary amine salts have the 1600 cm^{-1} band

Table 2.7 Miscellaneous R—H

Group	Band	Remarks
—S—H	2600–2550(w)	Weaker than O—H and less affected by H bonding
>P—H	2440–2350(m)	Sharp
$\overset{\overset{O}{\|\|}}{>}$P—OH	2700–2560(m)	Associated OH
>B—H	2640–2200(s)	
>Si—H	2250–2090(m)	
R—D	1/1.37 times the corresponding R—H frequency	Useful when assigning R—H bands, deuteration leading to a known shift to lower frequency

2.10   Absorption frequencies of triple bonds and cumulated double bonds

Table 2.8   Triple bonds

Group	Band	Remarks
—C≡C—H	3300(m) 2140–2100(w)	C—H stretching C≡C stretching
—C≡C—	2260–2150(v)	† (See Fig. 2.9)
—C≡N	2260–2200(v)	C≡N stretching; stronger and to the lower end of the range when conjugated; occasionally very weak (see Fig. 2.14) or absent; for example, some cyanohydrins show no C≡N absorption
Isonitriles R—N̄≡C̄	2150–2110(s)	
Nitrile oxides RC≡N̄—Ō	2305–2280	
Diazonium salts R—N≡N	∼2260	
Thiocyanates RS—C≡N	2175–2140(s)	Aryl thiocyanates at upper end of the range, alkyl at the lower end

† Conjugation with olefinic or acetylenic groups lowers the frequency and raises the intensity. Conjugation with carbonyl groups usually has little effect on the position of absorption.

Symmetrical and nearly symmetrical substitution makes the C≡C stretching frequency inactive in the infrared. It is, however, seen clearly in the Raman spectrum.

When more than one acetylenic linkage is present, and sometimes when there is only one, there are frequently more absorption bands in this region than there are triple bonds to account for them.

The ranges quoted in Table 2.9 are tentative, since relatively few compounds in some of these classes have been examined.

The usually high double bond frequencies encountered in the X=Y=Z systems are believed to arise from strong coupling of the two separate stretching vibrations, the asymmetrical and symmetrical stretching frequencies becoming widely separated. This type of coupling occurs only when two groups with similar high frequency vibrations and the same symmetry are situated near one another. Other examples in which such coupling is found are the amide group (Table 2.10) and the carboxylate ion (Table 2.10).

Table 2.9   Cumulated double bonds

Group	Band	Remarks
Carbon dioxide O=C=O	2349(s)	Appears in many spectra due to inequalities in path length
Isocyanates —N=C=O	2275–2250(s)	Very high intensity; position unaffected by conjugation
Azides —N$_3$	2160–2120(s)	
Carbodiimides —N=C=N—	2155–2130(s)	Very high intensity; split into an unsymmetrical doublet by conjugation with aryl groups
Ketenes $>$C=C=O	2155–2130(vs)	
Isothiocyanates —N=C=S	2140–1990(s)	Broad and very intense
Diazoalkanes R$_2$C=$\overset{+}{N}$=$\overset{-}{N}$	2050–2010(vs)	
Diazoketones RCOCH=$\overset{+}{N}$=$\overset{-}{N}$	2100–2050(s)	
Ketenimines C=C=N—	2050–2000(vs)	
Allenes C=C=C	1950–1930(s)	Two bands when terminal allene or when bonded to electron attracting groups, e.g. —CO$_2$H

2.11   The aromatic overtone and combination region, 2000–1600 cm^{-1}

Aromatic compounds are characterized by the weak C—H stretching band near 3030 cm^{-1} (Table 2.3) and by bands near 1600 and 1500 cm^{-1} (Table 2.14). Occasionally, the substitution pattern on a benzene ring can be deduced from the strong bands associated with the C—H out-of-plane bending vibrations below 900 cm^{-1} (see Table 2.15). In addition to these bands, there are bands in the 1225–950 cm^{-1} region which are of little use, and a group of weak overtone and combination bands in the 2000–1600 cm^{-1} region. It has been found from spectra taken with more concentrated solutions that the shape and number of the two to six bands found in this region is a function of the substitution pattern of the benzene ring. The use of this region, however, depends on the absence of other absorption in the region. The characteristic patterns for the various substituted benzenes are given in Nakanishi's book.

2.12   Absorption frequencies of the double bond region

Table 2.10   Carbonyl absorption $\diagdown$C$=$O *All bands quoted are strong.*

Groups	Band	Remarks
Acid anhydrides —CO—O—CO—		
Saturated	1850–1800 1790–1740	Two bands usually separated by about 60 cm^{-1}. The higher frequency band is more intense in acyclic anhydrides and the lower frequency band is more intense in cyclic anhydrides
Aryl and $\alpha\beta$-unsaturated	1830–1780 1770–1710	
Saturated five-ring	1870–1820 1800–1750	
All classes	1300–1050	One or two strong bands due to C—O stretching
Acid chlorides —COCl		
Saturated	1815–1790	Acid fluorides higher, bromides and iodides lower
Aryl and $\alpha\beta$-unsaturated	1790–1750	
Acid peroxides —CO—O—O—CO—		
Saturated	1820–1810 1800–1780	
Aryl and $\alpha\beta$-unsaturated	1805–1780 1785–1755	
Esters and lactones —CO—O—		
Saturated	1750–1735	
Aryl and $\alpha\beta$-unsaturated Aryl and vinyl esters	1730–1715	
C$=$C—O—CO—Alkyl	1800–1750	The C$=$C stretching band also shifts to higher frequency
Esters with electronegative α-substituents; e.g. $\diagdown$CCl—CO—O—	1770–1745	
α-keto esters	1755–1740	
Six-ring and larger lactones	Similar values to the corresponding open chain esters	
Five-ring lactone	1780–1760	
$\alpha\beta$-unsaturated five-ring lactone	1770–1740	When α-C—H present there are two bands, the relative intensity depending on the solvent (see Fig. 2.14)

Table 2.10 *continued*

Groups	Band	Remarks
$\beta\gamma$-unsaturated five-ring lactone; i.e. vinyl ester type	~1800	
Four-ring lactone	~1820	
β-keto ester in H-bonding enol form	~1650	Keto form normal; chelate-type H bond causes shift to lower frequency than the normal ester. The C=C is usually near 1630(s) cm^{-1}
All classes	1300–1050	Usually two strong bands due to C—O stretching

Aldehydes   —CHO
(see also Table 2.2 for C—H).
All values given below are lowered in liquid film or solid state spectra by about 10–20 cm^{-1}. Vapour phase spectra have values raised about 20 cm^{-1}

Saturated	1740–1720	
Aryl	1715–1695	(See Fig. 2.13a.) *Ortho* hydroxy or amino groups shift this value to 1655–1625 cm^{-1} due to intramolecular H bonding
α-chloro or bromo	1765–1730	
$\alpha\beta$-unsaturated	1705–1680	
$\alpha\beta,\gamma\delta$-unsaturated	1680–1660	
β-ketoaldehyde in enol form	1670–1645	Lowering caused by chelate-type H bonding

Ketones   C=O
All values given below are lowered in liquid film or solid state spectra by about 10–20 cm^{-1}. Vapour phase spectra have values raised about 20 cm^{-1}

Saturated	1725–1705	Branching at α position lowers frequency
Aryl	1700–1680	
$\alpha\beta$-unsaturated	1685–1665	Often two bands
$\alpha\beta,\alpha'\beta'$-unsaturated and diaryl	1670–1660	
Cyclopropyl	1705–1685	
Six-ring ketones and larger	Similar values to the corresponding open chain ketones	(See Fig. 2.8)
Five-ring ketones	1750–1740	$\alpha\beta$-unsaturation, etc. has a similar effect
Four-ring ketones	~1780	on these values as on those of open chain ketones

Table 2.10 *continued*

Groups	Band	Remarks
α-chloro or α-bromo ketones	1745–1725	Affected by conformation; highest values are obtained when both halogens are in the same plane as the C=O. α-F has larger effect, α-I has no effect
α,α'-dichloro or α,α'-dibromo ketones	1765–1745	
1,2-Diketones s-*trans* (i.e. open chains)	1730–1710 –	Antisymmetrical stretching frequency of both C=O's. The symmetrical stretching is inactive in the infrared but active in the Raman
1,2-Diketones s-*cis*, six-ring	1760 and 1730	
1,2-Diketones s-*cis*, five-ring	1775 and 1760	
Enolized 1,3-diketones	1650 and 1615	CO lowered by H bonding and C=C
o-Amino- or o-hydroxy-aryl ketones	1655–1635	Low due to intramolecular H bonding. Other substituents and steric hindrance, etc. affect the position of the band
Diazoketones	1645–1615	
Quinones	1690–1660	C=C usually near 1600(s) cm^{-1}
Extended quinones	1655–1635	
Tropone	1650	Near 1600 cm^{-1} when lowered by H bonding as in tropolones
Carboxylic acids —CO$_2$H		
All types	3000–2500	O—H stretching; a characteristic group of small bands due to combination bands, etc. (For the appearance of this group see Fig. 2.9)
Saturated	1725–1700	The monomer is near 1760 cm^{-1} but is rarely observed. Occasionally both bands, the free monomer and the H-bonded dimer can be seen in solution spectra. Ether solvents give one band near 1730 cm^{-1}
αβ-unsaturated	1715–1690	(See Fig. 2.9)
Aryl	1700–1680	
α-halo-	1740–1720	
Carboxylate ions —CO$_2^-$ For amino acids, see text below Fig. 2.10		
Most types	1610–1550 1420–1300	Antisymmetrical and symmetrical stretching respectively

Table 2.10 *continued*

Groups	Band	Remarks
Amides —CO—N$<$	(See also Table 2.5 and 2.6 for N—H stretching and bending)	
Primary —CONH$_2$		
In solution	~1690	Amide I; C=O stretching
Solid state	~1650	
In solution	~1600	Amide II; mostly N—H bending
Solid state	~1640	
		Amide I is generally more intense than amide II. (In the solid state amide I and II may overlap.) (See Fig. 2.11)
Secondary —CONH—		
In solution	1700–1670	Amide I (see Fig. 2.13)
Solid state	1680–1630	
In solution	1550–1510	Amide II; found in open chain amides only (see Fig. 2.13)
Solid state	1570–1515	Amide I is generally more intense than amide II
Tertiary	1670–1630	Since H bonding is absent solid and solution spectra are much the same (see Fig. 2.12)
Lactams		
Six- and larger rings	~1670	
Five-ring	~1700	Shifted to higher frequency when the N atom is in a bridged system
Four-ring	~1745	
R—CO—N—C=C		Shifted +15 cm^{-1} by the additional double bond
C=C—CO—N		Shifted by up to +15 cm^{-1} by the additional double bond. This is an unusual effect for $\alpha\beta$-unsaturation. It is said to be due to the inductive effect of the C=C on the well-conjugated CO—N system, the usual conjugation effect being less important in such a system
Imides —CO—N—CO—		
Cyclic six-ring	~1710 and ~1700	Shift of +15 cm^{-1} with $\alpha\beta$-unsaturation
Cyclic five-ring	~1770 and ~1700	
Ureas N—CO—N		
RNHCONHR	~1660	
Six-ring	~1640	
Five-ring	~1720	
Urethanes		
R—O—CO—N	1740–1690	Also shows amide II band when non- or mono-substituted on N
R—S—CO—N	1700–1670	

Table 2.10 *continued*

Groups	Band	Remarks
Thioesters and acids		
RCO—S—R'		
RCOSH	~ 1720	$\alpha\beta$-unsaturated or aryl acid or ester shifted ~ -25 cm^{-1}
RCOS—alkyl	~ 1690	
RCOS—aryl	~ 1710	
Carbonates		
RO—CO—Cl	~ 1780	
RO—CO—OR	~ 1740	
ArO—CO—OAr	~ 1785	
Cyclic 5-ring	~ 1820	
RS—CO—SR	~ 1645	
ArS—CO—SAr	~ 1715	

Intensities of carbonyl bands. Acids generally absorb more strongly than esters, and esters more strongly than ketones or aldehydes. Amide absorption is usually similar in intensity to that of ketones but is subject to much greater variations.

Position of carbonyl absorption. The general trends of structural variation on the position of C=O stretching frequencies may be summarized as follows:

1. The more electronegative the group X in the system R—CO—X—, the higher is the frequency.
2. $\alpha\beta$-unsaturation causes a lowering of frequency of 15–40 cm^{-1}, except in amides, where little shift is observed and that usually to higher frequency.
3. Further conjugation has relatively little effect.
4. Ring strain in cyclic compounds causes a relatively large shift to higher frequency. This phenomenon provides a remarkably reliable test of ring size, distinguishing clearly between four, five, and larger membered ring ketones, lactones, and lactams. Six-ring and larger ketones, etc. show the normal frequency found for the open chain compounds.
5. Hydrogen bonding to a carbonyl group causes a shift to lower frequency of 40–60 cm^{-1}. Acids, amides, enolized β-keto carbonyl systems, and *o*-hydroxy- and *o*-aminophenyl carbonyl compounds show this effect. All carbonyl compounds tend to give slightly lower values for the carbonyl stretching frequency in the solid state compared with the value for dilute solutions.
6. Where more than one of the structural influences on a particular carbonyl group is operating, the net effect is usually close to additive.

The most substituted double bonds tend to absorb at the high frequency end of the range, the least substituted at the low frequency end. The absorption may be very weak when the double bond is more or less symmetrically substituted, but the vibration frequency can then be detected and measured in the Raman spectrum. For the same reason *trans*-double bonds tend to absorb less strongly than *cis*-double bonds.

Table 2.11   Imines, oximes, etc. $>$C$=$N—

Group	Band	Remarks
$>$C$=$N—H	3400–3300(m)	N—H stretching; lowered on H bonding
$>$C$=$N—	1690–1640(v)	Difficult to identify due to large variations in intensity and the closeness to C$=$C stretching region. Oximes usually give very weak bands
$\alpha\beta$-unsaturated	1660–1630(v)	
Conjugated cyclic systems	1660–1480(v)	

Table 2.12   Azo compounds —N$=$N—

Group	Band	Remarks
—N$=$N—	1500–1400(v)	Very weak or inactive in infrared. Sometimes seen in Raman
$-\overset{+}{\underset{\underset{O^-}{\vert}}{N}}$=N—	1480–1450 1335–1315	Asymmetric and symmetric stretching

Table 2.13   Alkenes $>$C$=$C$<$

(See also Table 2.3 for the $=$C—H absorptions of alkenes.)

Group	Band	Remarks
Non-conjugated		
$>$C$=$C$<$	1680–1620(v)	May be very weak if more or less symmetrically substituted (see Fig. 2.12)
Conjugated with aromatic ring	~1625(m)	More intense than with unconjugated double bonds
Dienes, trienes, etc.	1650(s) and 1600(s)	Lower frequency band usually more intense and may hide or overlap the higher frequency band
$\alpha\beta$-unsaturated carbonyl compounds	1640–1590(s)	Usually much weaker than the C$=$O band (see, however, Fig. 2.11)
Enol esters, enol ethers and enamines	1690–1650(s)	(See Fig. 2.14)

Table 2.3 should be consulted for the $=$C—H vibration frequencies, which may give additional structural information.

A general trend which has been observed is the effect caused by strain on the stretching frequency of double bonds. A double bond exocyclic to a ring shows the same pattern as cyclic ketones: that is, the frequency rises as the ring size decreases.

Table 2.14 Aromatic compounds
(See also Tables 2.3 and 2.15 for aryl—H vibration frequencies.)

Group	Band	Remarks
Aromatic rings	~1600(m)	(See Fig. 2.13)
	~1580(m)	Stronger when the ring is further conjugated
	~1500(m)	This is usually the strongest of the two or three bands

A double bond within a ring shows the opposite trend: that is, the frequency falls as the ring size decreases. The C—H stretching frequency rises slightly as ring strain increases.

The two or three bands in the 1600–1500 cm^{-1} region are shown by most six-membered aromatic ring systems such as benzenes, polycyclic systems and pyridines. They constitute a valuable identification of such a system. Further bands are shown by aromatic rings in the fingerprint region between 1225 and 950 cm^{-1} which are of little diagnostic value. The weak overtone and combination bands in the 2000–1660 cm^{-1} region have been mentioned on page 46. A fourth group of bands below 900 cm^{-1} is produced by the out-of-plane C—H bending vibrations (see Table 2.15).

Table 2.15 Substitution patterns of the benzene ring

Group	Band	Remarks
Five adjacent H	770–730(s) and 720–680(s)	Monosubstituted
Four adjacent H	770–735(s)	*Ortho*-disubstituted (see Fig. 2.10)
Three adjacent H	810–750(s)	*Meta*-disubstituted, etc. and 1,2,3-trisubstituted
Two adjacent H	860–800(s)	*Para*-disubstituted, etc. (see Fig. 2.13a)
Isolated H	900–800(w)	*Meta*-disubstituted, etc.; usually not strong enough to be useful

The frequency of the C—H out-of-plane vibration is determined by the number of adjacent hydrogen atoms on the ring and hence the frequency is a means of determining the substitution pattern. This does not work as well in practice as one might hope. These strong bands are not always the only—or even the strongest—bands in the region (for example, *C*-halogen frequencies interfere particularly) so that assignments based on this evidence alone should be treated with caution. For example, the spectrum of tryptophan (Fig. 2.10) and the spectrum of *p*-acetamidobenzaldehyde (Fig. 2.13a) show only the characteristic absorption of *ortho*- and *para*-disubstituted benzene rings respectively; but the spectrum of *p*-nitrophenylpropiolic acid (Fig. 2.9) shows not only a band at 860 cm^{-1}, consistent with its being a *p*-disubstituted benzene, but also bands at 750 and 685 cm^{-1}, consistent with its being monosubstituted. It is obvious that a positive assignment of substitution pattern is not possible in the latter case.

The values in Table 2.15 hold reasonably well for condensed ring systems and for pyridines. Powerful electron withdrawing substituents tend to shift the values to higher frequency.

Table 2.16   Nitro, nitroso, etc. N=O

Group	Band	Remarks
C—NO$_2$	1570–1540(s) 1390–1340(s)	Lowered ~ 30 cm^{-1} when conjugated. The two bands are due to asymmetrical and symmetrical stretching of the NO bonds (see Fig. 2.9)
Nitrates O—NO$_2$	1650–1600(s) 1270–1250(s)	
Nitramines N—NO$_2$	1630–1550(s) 1300–1250(2)	
C—N=O	1600–1500(s)	Saturated aryl 1585–1540 1510–1490
O—N=O	1680–1610(s)	Two bands
N—N=O	1500–1430(s)	
$\overset{+}{>}$N$-\overset{-}{O}$ Aromatic Aliphatic	1300–1200(s) 970–950(s)	Very strong bands
NO$_3^-$	1410–1340 860–800	

2.13   Groups absorbing in the fingerprint region

Table 2.17   Ethers

Group	Band	Remarks
>C—O—C<	1150–1070(s)	C—O stretching
=C—O—C<	1275–1200(s) 1075–1020(s)	
C—O—CH$_3$	2850–2810(m)	C—H stretching; aryl ethers at higher end of the range
>C——C< \\O/	$\sim$1250 $\sim$900 $\sim$800	

Table 2.18 Silicon compounds

Group	Band	Remarks
Si—H	2250–2090(m) 1010–700(s)	Broad
$SiMe_n$	1275–1260(s) 880–760	$n=1 \sim 765$ $n=2 \sim 855$ $n=3 \sim 840$
Si—O	1110–1000 900–600	
Si—F	980–820	

Table 2.19 Sulphur compounds

Group	Band	Remarks
S—H	2600–2550(w)	S—H stretching; weaker than O—H and less affected by H bonding. This absorption is strong in the Raman
$\rangle$C=S	1200–1050(s)	
$\overset{}{\underset{\underset{S}{\parallel}}{C}}$—N$\langle$	~ 3400	N—H stretching; lowered to ~ 3150 cm^{-1} in the solid state
	1550–1450(s) 1300–1100(s)	Amide II Amide I
—O—CS—O	~ 1225	
—O—CS—N$\langle$	~ 1170	
$\rangle$N—CS—N$\langle$	1340–1130	
—S—CS—N$\langle$	~ 1050	
—S—CS—S—	~ 1070	
$\rangle$S=O	1060–1040(s)	
$\rangle$SO$_2$	1350–1310(s) 1160–1120(s)	
—SO$_2$—N$\langle$	1370–1330(s) 1180–1160(s)	
—SO$_2$—O—	1420–1330(s) 1200–1145(s)	

Table 2.19 *continued*

Group	Band	Remarks
—SO$_2$—Cl	1410–1375 1205–1170	
—S—F	815–755	

Table 2.20   Phosphorus compounds

Group	Band	Remarks
P—H	2440–2350(s)	Sharp
P—Ph	1440(s)	Sharp
P—O-alkyl	1050–1030(s)	
P—O-aryl	1240–1190(s)	
P=O	1300–1250(s)	
P=S	750–580	
P—O—P	970–910	Broad
$\overset{\overset{\displaystyle O}{\|}}{\diagup}$POH	2700–2560	H-bonded O—H
	1240–1180(s)	P=O stretching
P—F	1110–760	

Table 2.21   Boron compounds

Group	Band	Remarks
B—H	2640–2200(s)	
B—O	1380–1310(vs)	
B—N	1550–1330(vs)	
B—C	1240–620(s)	

Table 2.22   Halogen compounds

Group	Band	Remarks
C—F	1400–1000(s) 780–680	Sharp Weaker bands
C—Cl	800–600(s)	
C—Br	750–500(s)	
C—I	~500(s)	

Table 2.23   Inorganic ions

Group	Band	Remarks
Ammonium	3300–3030	All bands strong
Cyanide, thiocyanate, cyanate	2200–2000	
Carbonate	1450–1410	
Sulphate	1130–1080	
Nitrate	1380–1350	
Nitrite	1250–1230	
Phosphates	1100–1000	

2.14   Examples of infrared spectra

The following spectra show the appearance and relative intensities of the absorption peaks due to a number of functional groups. The wide variety of fingerprints shows the usefulness of this region for identification.

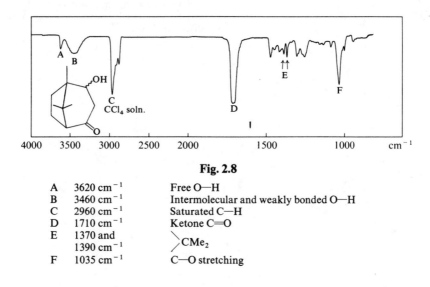

Fig. 2.8

A	3620 cm^{-1}	Free O—H
B	3460 cm^{-1}	Intermolecular and weakly bonded O—H
C	2960 cm^{-1}	Saturated C—H
D	1710 cm^{-1}	Ketone C=O
E	1370 and 1390 cm^{-1}	$\diagup\!\!\diagdown$CMe$_2$
F	1035 cm^{-1}	C—O stretching

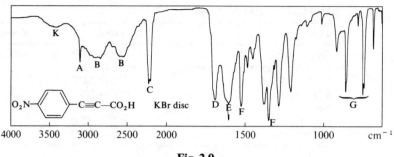

Fig. 2.9

A	3100 cm^{-1}	Aryl C—H stretching
B	3200–2400 cm^{-1}	Characteristic strongly H-bonded O—H of carboxylic acid
C	2225 cm^{-1}	Conjugated C≡C, hence strong
D	1690 cm^{-1}	Conjugated —CO$_2$H
E	1605 cm^{-1}	Benzene ring, unusually broad and unresolved. A band near 1500 cm^{-1} is masked
F	1520 and 1350 cm^{-1}	Conjugated nitro —NO$_2$
G	950–650 cm^{-1}	An example of a substituted ring in which it is not possible to decide with any certainty, due to the large number of bands in the region, in favour of 1,4-disubstitution. For an example where the assignment can be made with confidence see Fig. 2.10
K		OH of water almost always present even in good KBr discs like this one

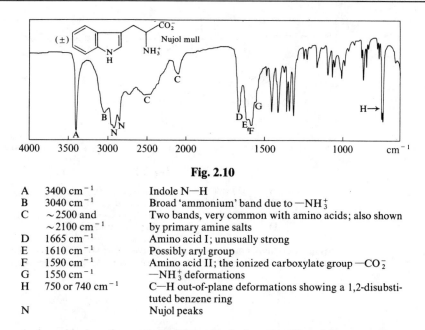

Fig. 2.10

A	3400 cm^{-1}	Indole N—H
B	3040 cm^{-1}	Broad 'ammonium' band due to —NH$_3^+$
C	~2500 and ~2100 cm^{-1}	Two bands, very common with amino acids; also shown by primary amine salts
D	1665 cm^{-1}	Amino acid I; unusually strong
E	1610 cm^{-1}	Possibly aryl group
F	1590 cm^{-1}	Amino acid II; the ionized carboxylate group —CO$_2^-$
G	1550 cm^{-1}	—NH$_3^+$ deformations
H	750 or 740 cm^{-1}	C—H out-of-plane deformations showing a 1,2-disubstituted benzene ring
N		Nujol peaks

Amino acids show the spectrum of the zwitterionic groups. The primary ammonium —NH$_3^+$ group N—H stretching appears under the peaks of the saturated C—H absorption. The two bands near 2500 and 2000 cm^{-1} are frequently found when the —NH$_3^+$ group is present, and are due to overtones and combinations. In the double bond region, there are several peaks, including at least one due to N—H bending and one, the strongest, due to the ionized carboxyl group. The highest frequency N—H bending peak is often very weak.

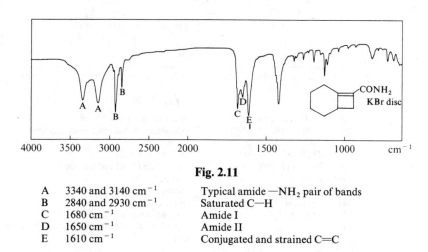

Fig. 2.11

A	3340 and 3140 cm^{-1}	Typical amide —NH$_2$ pair of bands
B	2840 and 2930 cm^{-1}	Saturated C—H
C	1680 cm^{-1}	Amide I
D	1650 cm^{-1}	Amide II
E	1610 cm^{-1}	Conjugated and strained C=C

This spectrum shows the pair of N—H stretching bands and the pair of bands in the C=O region typical of primary amides in the solid state. The C=C peak appears here as an unusually strong peak. The amide I and II bands are not always so well resolved in the solid state.

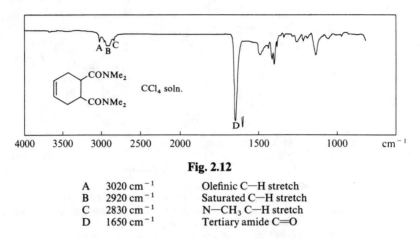

Fig. 2.12

A	3020 cm^{-1}	Olefinic C—H stretch
B	2920 cm^{-1}	Saturated C—H stretch
C	2830 cm^{-1}	N—CH$_3$ C—H stretch
D	1650 cm^{-1}	Tertiary amide C=O

This spectrum shows the absence of N—H, the strong sharp C=O of a tertiary amide, and, because of the symmetry of the molecule, no C=C stretching absorption.

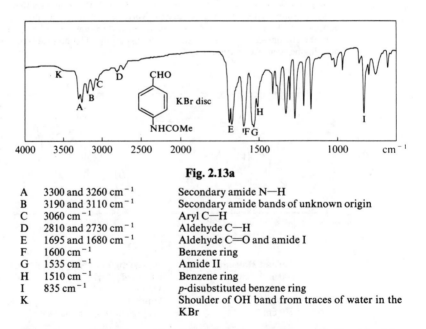

Fig. 2.13a

A	3300 and 3260 cm^{-1}	Secondary amide N—H
B	3190 and 3110 cm^{-1}	Secondary amide bands of unknown origin
C	3060 cm^{-1}	Aryl C—H
D	2810 and 2730 cm^{-1}	Aldehyde C—H
E	1695 and 1680 cm^{-1}	Aldehyde C=O and amide I
F	1600 cm^{-1}	Benzene ring
G	1535 cm^{-1}	Amide II
H	1510 cm^{-1}	Benzene ring
I	835 cm^{-1}	p-disubstituted benzene ring
K		Shoulder of OH band from traces of water in the KBr

This spectrum shows the multiplicity of bands found with secondary amides. The presence of so many bands in the spectra of such compounds as secondary amides is probably caused by the many ways in which such groupings can associate with each other, of which those shown on page 43 are only two of many possibilities.

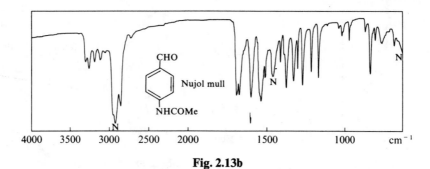

Fig. 2.13b

This spectrum was taken on the same compound as that in Fig. 2.13a, but was done as a Nujol mull. The main features are closely similar, but it can be seen how one of the Nujol peaks (labelled N) can obscure an important peak such as one of those labelled D on Fig. 2.13a. On the other hand, the band marked K on Fig. 2.13a is no longer present.

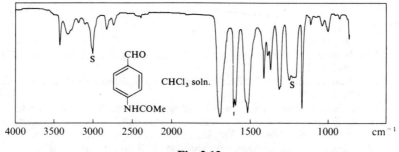

Fig. 2.13c

This is again a spectrum of the same compound, but taken in solution. This time some changes in appearance are apparent. The N—H stretch region is markedly different, and the amide I band has moved to a slightly higher frequency, making it coincident with the aldehyde C=O band. These changes are expected of a secondary amide, where a change in the nature of the intermolecular associations occurs in going from the solid state into the solution state. Such changes most affect the vibration frequencies of the functional groups involved in those associations.

The benzene ring band near 1600 cm^{-1} is now resolved into the two bands often found here: solution state spectra are often better resolved than solid state spectra. On the other hand there are many more strong lines in the fingerprint region in the solid state spectra.

The bands marked S are partly due to the solvent, because the absorption of the solvent has been incompletely cancelled (see page 30).

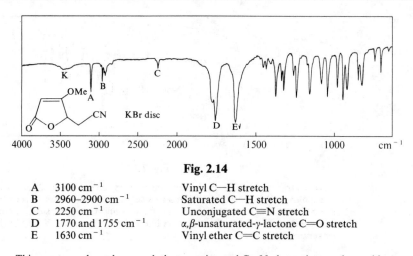

Fig. 2.14

A	3100 cm^{-1}	Vinyl C—H stretch
B	2960–2900 cm^{-1}	Saturated C—H stretch
C	2250 cm^{-1}	Unconjugated C≡N stretch
D	1770 and 1755 cm^{-1}	α,β-unsaturated-γ-lactone C=O stretch
E	1630 cm^{-1}	Vinyl ether C=C stretch

This spectrum shows how weak the unconjugated C≡N absorption can be, and how αβ-unsaturation, which lowers the frequency of the vibrations of the carbonyl group, combines with the presence of a five-membered ring, which raises the frequency, to give a band at 1755 cm^{-1}, near the normal position of saturated esters and six-membered ring lactones. The extra band at 1770 cm^{-1} is common with αβ-unsaturated five-membered ring lactones having an α—H.

Bibliography

TEXTBOOKS

L. J. Bellamy, *The Infrared Spectra of Complex Molecules*, Chapman and Hall, London. Vol. 1, 1975, Vol. 2, 1980.

J. P. Ferraro and L. J. Basile, *Fourier Transform Infrared Spectroscopy*, Academic Press, New York, 1978.

P. Griffiths, *Fourier Transform Infrared Spectrometry*, Wiley, New York, 2nd Ed., 1986.

A. R. Katritzky and A. P. Ambler, Chapter 10, Infrared Spectra, *Physical Methods in Heterocyclic Chemistry*, Vol. II, Academic Press, New York, 1963.

D. A. Long, *Raman Spectroscopy*, McGraw-Hill, Maidenhead, 1976.

K. Nakanishi, *Infrared Absorption Spectroscopy*, Holden-Day, San Francisco, 2nd Ed., 1977.

J. H. van der Maas, *Basic Infrared Spectroscopy*, Heyden, London, 2nd Ed., 1972.

THEORETICAL TREATMENTS

G. R. Barrow, *Introduction to Molecular Spectroscopy*, McGraw-Hill, New York, 1962.

P. R. Griffiths and J. A. de Haseth, *Fourier Transfer Infrared Spectroscopy*, Wiley, New York, 1986.

G. Herzberg, *Infrared and Raman Spectra of Polyatomic Molecules*, Van Nostrand, Princeton, 1945.

R. G. J. Miller and B. C. Stace, *Laboratory Methods in Infrared Spectroscopy*, Heyden, London, 2nd Ed., 1972.

CATALOGUES OF SPECTRA

The Aldrich Library of FT-IR Spectra, Aldrich Chemical Company, Milwaukee, 1985.

D. Dolphin and A. Wick, *Tabulation of Infrared Spectral Data*, Wiley, New York, 1977.

E. Pretsch, T. Clerc, J. Seibl, and W. Simon, *Tables of Spectral Data for Structure Determination of Organic Compounds*, Springer, Berlin, English Ed., 1983.

Sadtler Handbook of Infrared Grating Spectra, Heyden, London.

3. Nuclear magnetic resonance spectra

3.1 *Nuclear spin and resonance.* 3.2 *The measurement of spectra.* 3.3 *The chemical shift.* 3.4 *The intensity of NMR signals and integration.* 3.5 *Factors affecting the chemical shift.* 3.6 *Spin–spin coupling to* ^{13}C. 3.7 *^1H–^1H first-order coupling.* 3.8 *Some simple ^1H–^1H splitting patterns.* 3.9 *The magnitude of ^1H–^1H coupling constants.* 3.10 *Line broadening and environmental exchange.* 3.11 *Improving the NMR spectrum.* 3.12 *The many-pulse experiment: new techniques in FT NMR spectroscopy.* 3.13 *The separation of chemical shift and coupling onto different axes.* 3.14 *Spin decoupling.* 3.15 *The nuclear Overhauser effect.* 3.16 *Associating the signals from directly bonded ^{13}C and ^1H.* 3.17 *Identifying other connections.* *Tables of Data.* *Bibliography.*

3.1 Nuclear spin and resonance

The phenomenon of nuclear magnetic resonance was first observed in 1946, and it has been routinely applied in organic chemistry since about 1960. It has grown enormously in power and versatility since that time, conspicuously since the late 1970s with the introduction of Fourier transform (FT) NMR spectroscopy on a routine basis. The subject has grown so fast that it has almost become a scientific discipline in its own right, and we can introduce it here only in the most superficial way. Fortunately, the technique can still be applied to most problems of structure determination without the help of experts.

Some atomic nuclei have a nuclear spin (I), and the presence of a spin makes these nuclei behave rather like bar magnets. In the presence of an applied magnetic field the nuclear magnets can orient themselves in $2I + 1$ ways. Those nuclei with an odd number of nucleons, of which the most important are ^1H and ^{13}C, have spins of $\frac{1}{2}$. These nuclei, therefore, can take up one of only two orientations, a low energy orientation aligned with the applied field, and a high energy orientation opposed to the applied field. The difference in energy is given by:

$$\Delta E = h\gamma B_0/2\pi \qquad (3.1)$$

where γ is the magnetogyric ratio (a proportionality constant, differing for each nucleus, which essentially measures the strength of the nuclear magnets) and B_0 is the strength of the applied magnetic field. The number of nuclei in the low energy state (N_α) and the number in the high energy state (N_β) will differ by an amount determined by the Boltzmann distribution:

$$N_\beta/N_\alpha = \exp(-\Delta E/kT) \qquad (3.2)$$

When a radio frequency signal is applied to the system, this distribution is changed if the radio frequency matches the frequency at which the nuclear magnets naturally precess in the magnetic field B_0: some of the N_α nuclei are promoted from the low energy state to the high energy state, and N_β increases (Fig. 3.1). The frequency in Hz, the resonance frequency, is given by:

$$v = \gamma B_0/2\pi \tag{3.3}$$

and is therefore dependent upon both the applied field strength and the nature of the nucleus in question. The frequencies at which some commonly encountered nuclei come into resonance at a field strength of 2.35 Tesla (23.49 kilogauss) are given in Table 3.4, but for now it is enough to know that at this field strength ^{13}C comes into resonance at 25.14 MHz and ^1H at 100 MHz. Because of the very widespread use of proton NMR spectroscopy, it is common to refer to an instrument with this field strength as a 100 MHz instrument, even when it is measuring ^{13}C spectra. This field strength is attainable with an electromagnet, but higher field strengths, which have considerable advantages, are obtained using superconducting magnets cooled with liquid helium.

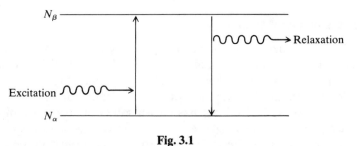

Fig. 3.1

The spectrum is measured on one of two ways. In the traditional continuous wave (CW) method, a small proportion of the applied signal is absorbed in promoting some of the N_α nuclei to the higher energy state, this response is picked up in a receiving coil. The frequency range being studied is scanned steadily from one extreme to the other, either by varying the frequency of the transmitter or by varying the magnetic field. The spectrum is then plotted directly, as it is being taken, as absorption (upwards) against frequency (increasing towards the left, as in UV and IR spectroscopy), the whole process taking a few minutes. Alternatively, the radio frequency signal is applied as a single powerful pulse, which effectively covers the whole frequency range and lasts for a time (t_p) typically of a few microseconds. This pulse generates an oscillating magnetic field (B_1) along the x axis, at right angles to the applied magnetic field (B_0) which is defined as being along the z axis (Fig. 3.2a). Because of the small difference between N_α and N_β, the sample being investigated has a net magnetization (M), which, because N_α is larger than N_β, is initially aligned in the direction of the applied field. The effect of the pulse is to tip the magnetization through an angle given by:

$$\Theta = \gamma B_1 t_p \tag{3.4}$$

Commonly, the time (t_p) is chosen so that Θ is $90°$, and such pulses are called $\pi/2$ pulses. The magnetization, disturbed from its orientation along the z axis, precesses in the xy plane (Fig. 3.2b), just as a gyroscope precesses when it is tipped out of the axis of the gravitational field. After the pulse is over, a receiver coil picks up the resultant oscillation during an acquisition period, typically of a few seconds. This signal is called the free induction decay (FID) and is a complicated wave pattern decaying away to zero during the acquisition period. The decay takes place because the individual nuclei relax, through interaction with local fluctuating magnetic fields (more on this later), back to their equilibrium states. Fourier transformation of the FID, which is said to be in the time domain, converts it into a spectrum, which is said to be in the frequency domain. The spectrum (an FT spectrum) is then plotted in exactly the same form as a CW spectrum.

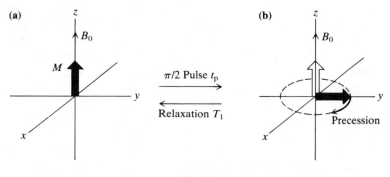

Fig. 3.2

Normally the difference between N_α and N_β is very small; for protons in a 200 MHz instrument it is only of the order of 1 in 10^5. Therefore, NMR spectroscopy is a relatively insensitive technique compared with UV and IR spectroscopy, and the electronics need to be very sensitive. The higher the field strength, B_0, the greater the difference between N_α and N_β (Eqs 3.1 and 3.2), which means that high-field instruments (200–500 MHz) are inherently more sensitive than the older (60–100 MHz) machines. FT-NMR has the further advantages that all nuclei of a given type can be excited simultaneously, and that large numbers of pulses can be applied, each followed by an acquisition period. Because the FID is handled in digital form in a computer, successive FIDs are easily added together, and the Fourier transformation can be carried out on the sum. Not only is the absorption intensified by this procedure but noise, which is random, is largely cancelled out. The accumulation of n spectra improves the signal to noise ratio by $\sqrt{n}$ relative to that obtained in a single spectrum. With nuclei such as ^{13}C, where this isotope is present in only one in a hundred carbon atoms and where the nucleus is inherently much less sensitive anyway, there is an even greater need for high-field machines using FT.

In organic chemistry, the most commonly encountered nuclei with $I = \pm\frac{1}{2}$ are ^1H, ^{13}C, ^{19}F, ^{29}Si, and ^{31}P. The common nuclei with $I = 0$, ^{12}C and ^{16}O, are completely inactive. A few other common nuclei have spins, of which ^2H and ^{14}N ($I = 1$) are perhaps the most important. When present in organic molecules, they affect ^{1}H and ^{13}C NMR spectra, but it is comparatively unusual to study the NMR spectra of these nuclei themselves.

3.2   The measurement of spectra

CW spectra are now taken only for ^{1}H spectra on samples of 50 mg or more, typically to give the first spectrum of a product from a reaction mixture. FT spectra are taken when the samples are smaller than this, when accurate and well-resolved spectra are needed, and for all ^{13}C spectra. The sample size for a routine ^{13}C spectrum is about 50–100 mg, and for a FT ^{1}H spectrum only 1–10 mg. However, with a suitable investment in a large number of pulses, it is possible to obtain high-quality ^{13}C spectra from 1 mg and ^{1}H spectra from less than 0.1 mg, if the molecular weight is not more than a few hundred. The sample is dissolved in a solvent, preferably one which does not itself give rise to signals in the NMR spectrum. The most commonly used solvents are CCl_4, $CDCl_3$, C_6D_6, d_6-DMSO [$(CD_3)_2SO$], and D_2O, the choice of solvent being largely determined by the solubility of the compound under investigation. The solution is introduced into a precision-ground tube—most commonly of 5 mm diameter—to a depth of 2–3 cm. The solution should be free both of paramagnetic and insoluble impurities, and it should not be viscous, or resolution may suffer. An internal reference compound (TMS, see below) is added to the solution, and the tube is lowered into a probe placed between the poles of the magnet. The probe has the transmitter and receiver coils connected to it. The magnet is then tuned to give the highest possible level of homogeneity, generally about 1 in 10^9, and the tube is spun (at about 30 r.p.s.) about its vertical axis to improve the effective homogeneity even further. The spectrum is then taken using the instrument controls—the knobs and switches of a CW instrument and the computer keyboard of an FT instrument—and is printed automatically using a pen and chartpaper. FT spectra are also usually available in digital form as a table of absorption peaks listed by frequency and intensity.

3.3   The chemical shift

So far the discussion has seemed to imply that an NMR spectrum consists of an absorption line from each kind of magnetically active nucleus present in the sample under study: one absorption peak from the ^{13}C atoms and one from the protons, for example. This, of course, would not be very informative.

In practice the range of frequencies looked at in any one spectrum is much narrower. In ^{13}C spectra—taken, for example, on a 100 MHz machine—the range of frequencies is a small segment of about 5000 Hz in the neighbourhood of the resonance frequency 25.14 MHz. This range is wide enough to bring each of the different ^{13}C atoms in most organic compounds successively into resonance. The precise frequency at which each carbon comes into resonance is determined not only by the applied field, B_0, but also by minute differences in the magnetic environment experienced by each nucleus. These minute differences are caused largely by the variation in electron density in the neighbourhood of each nucleus, with the result that each chemically distinct carbon atom in a structure, when it happens to be a ^{13}C, will come into resonance at a slightly different frequency from all the others. The electrons affect the microenvironment, because their movement creates a magnetic field.

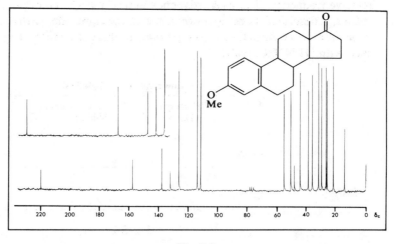

Fig. 3.3

Most machines routinely resolve signals only 0.5 Hz apart. With a spread of 5000 Hz, coincidence of ^{13}C signals is comparatively rare. Figure 3.3 shows the ^{13}C NMR spectrum of oestrone methyl ether. Each upward line corresponds to one of the 19 carbon atoms. Because the spread of frequencies is caused by the different chemical (and hence magnetic) environments, the signals are described as having a *chemical shift* from some standard frequency. In practice it is inconvenient to characterize the peaks by assigning to them a particular frequency, all very close to 25.14 MHz: the numbers are cumbersome and difficult to measure accurately. Furthermore they change from machine to machine, and even from day to day, as the applied field changes. It is convenient instead to measure the difference of the frequency (v_s) of the peak from some internal standard (both measured in Hz) and to divide this by the operating frequency in MHz to obtain a field-independent number in a convenient range. The internal standard almost always used is tetramethylsilane (TMS), and the chemical shift scale δ is then defined by:

$$\delta = v_s \text{ (Hz)} - v_{\text{TMS}} \text{ (Hz)/operating frequency (MHz)} \tag{3.5}$$

The parameter δ, which measures the position of the signal, will now be the same whatever machine it is measured on. It has no units and is expressed as fractions of the applied field in parts per million (p.p.m.). Tetramethylsilane is chosen as the internal standard because it is inert, volatile, non-toxic, and cheap, and it has only one signal, which comes into resonance at one extreme of the frequencies found for most carbon atoms in organic structures. The scale on Fig. 3.3. shows the common range of δ values, which are always written from right to left, a *convention* that denotes frequencies as higher and positive on the left and lower and positive on the right. The TMS signal, were it to be shown, would be on the extreme right on zero, by definition. Unfortunately, the left and right sides are hardly ever referred to as being at the high and low frequency end of the spectrum. Instead, for historical reasons, they are almost invariably referred to by the value of the applied field which corresponds to this relative frequency. The *high* frequency end of the spectrum, on the left with *high δ* values, is described as being *downfield*, and the right side, with *low δ* values is said to be *upfield*. It is essential that you get used to this convention (Fig. 3.4), which is also used with ^{1}H NMR spectra.

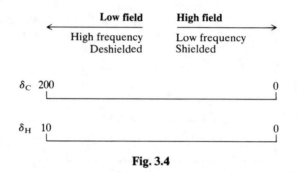

Fig. 3.4

In ^{1}H spectra, taken on a 100 MHz machine, a narrower range—about 1000 Hz— is enough to bring all the protons into resonance. Resolution is usually possible to 0.5 Hz, and it is therefore more common to find coincident signals from different protons than it is to find coincident signals from ^{13}C atoms. Resolution gets better with instruments working at higher field, but coincidence, or at any rate overlap, is still common. The same field-independent $δ$ scale, as defined by Eq. 3.5, is used in ^{1}H NMR spectroscopy, except that v_{TMS} is now the resonance frequency of the protons,

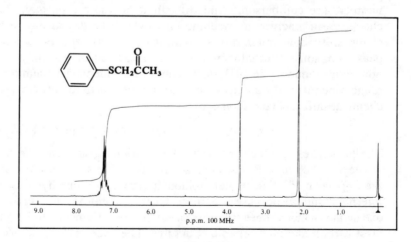

Fig. 3.5

rather than the carbon nuclei, in TMS. Thus the protons in most organic compounds come into resonance within 10 p.p.m. downfield of TMS.

Figure 3.5 shows the ^1H NMR spectrum of 1-phenylthiopropan-2-one. The TMS signal is visible at $\delta = 0$, and there are three absorptions corresponding to the three groups of protons, the aromatic protons in an ill-resolved group at low field, the methylene protons in the middle, and the methyl protons at high field.

3.4   The intensity of NMR signals and integration

The absorption of the signal in CW spectra is generally proportional to the number of protons coming into resonance at the frequency of the signal, with the result that the area under the absorption peaks is proportional to the number of protons being detected. This is shown on Fig. 3.5, where the machine has plotted an integration trace as a horizontal line rising from left to right as it passes each absorption. The extent of the rise measures the integration, and shows that the number of hydrogens in each peak is 5, 2, and 3, reading from left to right. (The actual numbers, normalized to add up to 10, are 5.00, 1.93, and 3.07, which gives some idea of how accurate a routine integration is.)

FT spectra are not quite so straightforward. For integration to be reliable, it is essential that all the nuclei relax to their equilibrium distribution between successive pulses. This is normally the case with properly taken ^1H spectra, but it is not the case with ^{13}C spectra. Figure 3.3 shows that the peaks are of markedly different intensity, yet each corresponds to a single, chemically distinct carbon atom. For relaxation to occur, the precessing magnetization of Fig. 3.2 must interact with local fluctuations in the magnetic fields in the molecule, especially those caused by other nuclear magnets. For this reason, the relaxation rate of carbon atoms directly bonded to hydrogen atoms is much higher than for carbon atoms not so bonded. This can be seen in Fig. 3.3, where the five lowest-intensity peaks, at δ 47.8, 131.7, 137.4, 157.3, and 220.2, correspond to the five fully substituted carbons in the molecule. Integration cannot therefore be used in ^{13}C NMR spectroscopy in the way that it can in ^1H NMR spectroscopy, and it is not uncommon to find that some peaks are so weak (carbonyl groups are notorious in this respect) that they do not appear in the spectrum. It is possible to increase the intensity of weak signals by supplying a powerful magnetic influence, such as a paramagnetic salt, to speed up the relaxation. This is shown in the upper trace of Fig. 3.3., which reproduces, slightly displaced, the five downfield signals recorded after chromium acetylacetonate had been added to the solution: the four signals furthest downfield have all increased in intensity relative to the upfield signal (C-1).

3.5   Factors affecting the chemical shift

Intramolecular factors affecting the chemical shift
The inductive effect.   In a uniform magnetic field, the electrons surrounding a nucleus circulate, setting up a secondary magnetic field opposed to the applied field at the nucleus (Fig. 3.6a, where the solid lines indicate the lines of force associated with the

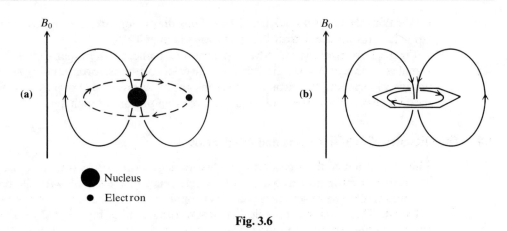

(a)

(b)

● Nucleus

• Electron

Fig. 3.6

induced field). As a result, nuclei in a region of high electron density experience a field proportionately weaker than those in a region of low electron density, and a higher field has to be applied to bring them into resonance. Such nuclei are said to be *shielded* by the electrons. In summary, a high electron density shields a nucleus, and causes resonance to occur at relatively high field (i.e. with low values of δ). Likewise, a low electron density causes resonance to occur at relatively low field (i.e. with high values of δ), and the nucleus is said to be *deshielded*. The extent of the effect can be seen in the positions of resonance of the ^{1}H and ^{13}C nuclei of the methyl group attached to the various atoms listed in Table 3.1. The electropositive elements (Li, Si) shift the signals upfield, and the electronegative elements (N, O, Cl) shift the signals downfield, because they donate and withdraw electrons, respectively.

Table 3.1 Chemical shifts for methyl groups attached to various atoms in CH$_3$X

X	δ_C	δ_H	X	δ_C	δ_H
Li	-14.0	-1.94	NH$_2$	26.9	2.47
SiMe$_3$	0.0	0.0	OH	50.2	3.39
H	-2.3	0.23	F	75.2	4.27
Me	8.4	0.86	SMe	19.3	2.09
Et	15.4	0.91	Cl	24.9	3.06

Hydrogen is more electropositive than carbon, with the result that every replacement of hydrogen by an alkyl group causes a downfield shift in the resonance of that carbon, and any remaining hydrogens on it. Thus methyl, methylene, methine, and quaternary carbons (and their attached hydrogens) come into resonance at successively lower field (**1–5**).

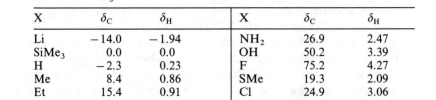

δ_C -2.3	δ_C 8.4	δ_C 15.9	δ_C 25.0	δ_C 27.7
CH$_4$	**MeCH$_3$**	**Me$_2$CH$_2$**	**Me$_3$CH**	**Me$_4$C**
δ_H 0.23	δ_H 0.86	δ_H 1.33	δ_H 1.68	
1	**2**	**3**	**4**	**5**

Anisotropy of chemical bonds. Chemical bonds are also regions of high electron density that can set up magnetic fields. These fields are stronger in one direction than another (they are anisotropic), and the effect of the field on the chemical shift of nearby nuclei is dependent upon the orientation of the nucleus in question with respect to the bond. π-bonds are especially effective in influencing the chemical shift of nearby atoms, so that allylic carbons and hydrogens are shifted downfield (to higher δ values) relative to their saturated counterparts, and the olefinic carbon atoms and the hydrogens bonded to them are shifted even further (Table 3.2), partly because of the anisotropy and partly because trigonal carbons are more electronegative (because of their higher s character) than tetrahedral carbon atoms. A carbonyl group has a similar effect on adjacent atoms (Table 3.2), and the carbonyl carbons themselves (and the hydrogens bonded to them) resonate at the lowest field position normally found in organic structures, since they suffer the combined effects of the induced anisotropic field and a nearby electronegative element.

Table 3.2 Chemical shifts of carbon and hydrogen in, on, and near multiple bonds

Compound	δ_C	δ_H	Compound	δ_C	δ_H
CH_3H	-2.3	0.23	CH_3CHO	31.2	2.20
$CH_3CH{=}CH_2$	22.4	1.71	CH_3COCH_3	28.1	2.09
$CH_3C{\equiv}CH$		1.80	CH_3CN	1.30	1.98
$CH_2{=}CH_2$	123.3	5.25	CH_3CHO	199.7	9.80
$CH_3C{\equiv}CH$	66.9	1.80	CH_3COCH_3	206.0	
$CH_3C{\equiv}CCH_3$	79.2		CH_3CN	117.7	

When a double bond carries a polar group, the electron distribution is displaced. The displacement is usually understood as a combination of inductive effects, which operate in the σ-framework (and simply fall off with distance) and conjugative effects, which operate in the π-system (and alternate along a conjugated chain). The effects in the π-system can be illustrated with curly arrows, as in the structures **6** and **7**, which have π-donor and π-acceptor groups, respectively.

6   **7**

These displacements of electron density naturally affect the position of resonance of nearby nuclei, as shown in Table 3.3. In general, π-donor groups shield the β nuclei, as implied by the curly arrows in **6**, causing an upfield shift relative to their position in ethylene. They simultaneously induce a downfield shift of the α nuclei. The π-acceptor groups ($COCH_3$ and $SiMe_3$) cause downfield shifts both of the β and α nuclei, with the effect at the β position large because of overlap (**7**, arrows).

Table 3.3 Conjugative effects on chemical shifts of substituted alkenes

$$\begin{array}{c} H_\beta \qquad H_\alpha \\ \diagdown \qquad \diagup \\ C_\beta{=}C_\alpha \\ \diagup \qquad \diagdown \\ H \qquad X \end{array}$$

X	Electronic nature	$\delta_{C\beta}$	$\delta_{C\alpha}$	$\delta_{H\beta}$	$\delta_{H\alpha}$
H	Reference compound	123.3	123.3	5.28	5.28
Me	Weak π- and σ-donor	115.4	133.9	4.88	5.73
OMe	π-donor, σ-acceptor	84.4	152.7	3.85	6.38
Cl	σ-acceptor, weak π-donor	117.2	125.9	5.02	5.94
CH=CH$_2$	Simple conjugation	130.3	136.9	5.06	6.27
SiMe$_3$	π-acceptor, σ-donor	129.6	138.7	5.87	6.12
COMe	π-acceptor, σ-acceptor	129.1	138.3	6.40	5.85

An important anisotropic effect is produced by the π-system of aromatic rings. The circulating electrons are called a ring current, and they create a magnetic field opposed to the applied field at the centre of the ring, but reinforcing the applied field outside the ring (Fig. 3.6b). The effect of the induced field is to deshield substantially the hydrogens attached to the aromatic ring **9**, which generally come into resonance 1.5–2.00 p.p.m. downfield from the corresponding olefinic signals **8**. The ^{13}C signal for benzene is similarly shifted downfield (by 5.2 p.p.m.), but the effect is much less noticeable because of the relatively large chemical shifts. In the special case of the aromatic [18]-annulene (**10**), there are 'inside' as well as 'outside' hydrogens. The 'inside' hydrogens experience a weaker field than the applied field, and come into resonance at a conspicuously high field, while the 'outside' hydrogens come into resonance in the normal aromatic region.

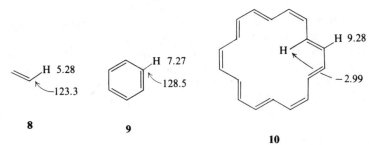

8   **9**   **10**

Polar groups attached directly to a benzene ring cause upfield and downfield shifts more or less in the same way that they do on a simple double bond. The effects of a π-donor and a π-acceptor group are seen in the structures **11** and **12**, where the signals of the *ortho* and *para* carbons and hydrogens are shifted upfield by the methoxy group, relative to the signals of benzene (**9**). The effect of the nitro group is less straightforward: the *ortho* hydrogen and the *para* carbon and hydrogen are shifted downfield, as one might expect, but the *ortho* carbon is shifted upfield.

Very approximately, a change in substitution pattern in any organic structure has a similar effect on both the carbon and the proton spectra: the δ value of the carbon signal is about 20 times the δ value of the proton signal. However, there are many

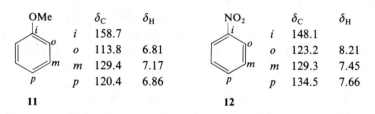

OMe		δ_C	δ_H		NO$_2$		δ_C	δ_H
	i	158.7				i	148.1	
	o	113.8	6.81			o	123.2	8.21
	m	129.4	7.17			m	129.3	7.45
	p	120.4	6.86			p	134.5	7.66
11					**12**			

large deviations from this general picture, including the opposite effect of the nitro group on the *ortho* carbons and hydrogens of **12**, and the big effect of the aromatic ring current on protons compared with the appearance of a small effect on carbons. The overall pattern that has emerged from the discussion so far can be summarized in Fig. 3.7.

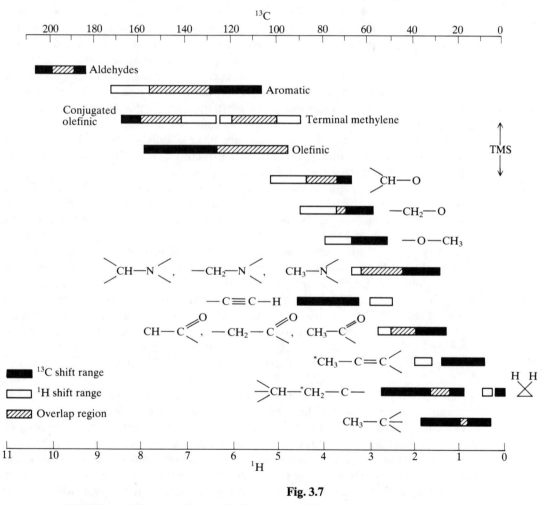

Fig. 3.7

(Reproduced with permission from D. Shaw, *Fourier Transform NMR Spectroscopy*, Elsevier, Amsterdam, 1976, p. 235.)

One curious feature in Fig. 3.7 has not been mentioned: the carbon and proton resonances of cyclopropanes appear at exceptionally high field, above the usual range for methylene groups, and even above the range for most methyl groups. In cyclopropane itself, the three *cis* vicinal C—H bonds are able to conjugate with each other, just as p orbitals conjugate with each other. The cyclic six-electron conjugated system thus set up gives rise to a ring current, and both the protons and the carbons sit in the shielding region of the magnetic field induced by that ring current. They come into resonance at δ_C −2.8 and δ_H 0.22. Substitution by alkyl groups and by electronegative elements moves the resonances downfield in the usual way.

Estimating a chemical shift.   The inductive and anisotropic effects are more or less additive, so that we can, for example, see that the signals in the spectrum of Fig. 3.5 are in appropriate places. The COMe group (at δ 2.15) and the phenyl group (as a multiplet at 7.1–7.4) are simply in the appropriate regions (see Table 3.2 and **9**). The methylene group (at 3.69) is also in an appropriate region for the combined effect of its being adjacent both to a carbonyl group and a sulphur; we can take δ 0.23 as a starting point, the position of resonance of methane's protons; we add 1.86 for the effect of a sulphur group (in Table 3.1, the sulphur group causes a downfield shift of 1.86 p.p.m. to a methyl group) and, coincidentally, 1.86 for the effect of the carbonyl group (in Table 3.2, the carbonyl group also causes a downfield shift of 1.86). The calculated value using this simple procedure (0.23 + 1.86 + 1.86) is 3.95, which is close to the observed value. The additivity principle has its limitations, but gives results usually within 0.6 p.p.m. of the observed values.

There are several sets of empirical rules of this kind, to deal with the different environments in which ^{13}C and ^1H are commonly encountered. These rules, and the tables of data needed to apply them, are all grouped together at the end of this chapter, where they can be easily found when you use this book as a handbook in the laboratory. Using them, you can estimate the chemical shift of the different kinds of carbon atoms in simple aliphatic compounds (Eq. 3.15 and Tables 3.6, 3.7, and 3.8), in simple alkenes (Eq. 3.16 and Table 3.10), and in polysubstituted benzene rings (Eq. 3.17 and Table 3.12), and of the carbon atoms in the various kinds of carbonyl groups (Table 3.13). There are similar rules and tables, more accurate than the data used in the example above, but based on the same idea, for estimating the proton chemical shifts of substituted alkanes (Eq. 3.19 and Tables 3.17, 3.18, and 3.19), substituted alkenes (Eq. 3.20 and Tables 3.20, 3.21, and 3.22), and substituted benzene rings (Eq. 3.21 and Tables 3.22 and 3.23).

As we have already seen, the estimation of chemical shift cannot be expected to give very accurate answers. The conformation of one molecule may not be the same as the conformation of the model on which the rules were based; the anisotropy of the field then causes the local field in the compound under investigation to differ from that of the model. A second reason for being cautious when interpreting chemical shift in detail is that the effects of distant groups are not included in the rules, usually because they are relatively unimportant. In some molecules, however, a distant group may fold back into the region of the nucleus under investigation, and shift its resonance dramatically. This is especially the case when aromatic rings are present.

Intermolecular factors affecting the chemical shift

Hydrogen bonds. A hydrogen atom involved in hydrogen bonding is sharing its electrons with two electronegative elements. As a result, it is itself deshielded and comes into resonance at low field. In water, in very dilute solution in $CDCl_3$, hydrogen bonding is at a minimum for an OH group, and the protons come into resonance at δ ~1.5. In droplets of water, on the other hand, suspended in $CDCl_3$, the molecules are hydrogen bonded intermolecularly, and they come into resonance at δ ~4.8. These signals can sometimes be seen in FT spectra taken on very small samples. The position of resonance of the OH and NH protons of alcohols and amines is unpredictable, because the extent to which the hydrogen atoms are involved in hydrogen bonding is both unpredictable and concentration dependent. The usual range is δ 0.5–4.5 for alcohols and 1–5 for amines. Thiols, which have weaker hydrogen bonds, come into resonance in the narrower range δ 3–4. The much stronger intermolecular hydrogen bonding in carboxylic acid dimers **13**, leads to very low-field absorption in the δ 9–15 range, and the corresponding intramolecular hydrogen bonding of enolized β-diketones **14** is similar (δ 15.4 for **14** itself). These are off the scale of the usual ^{1}H NMR trace, and have to be looked for specially.

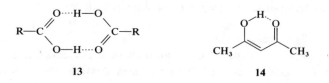

13   14

Fortunately it is easy to identify these kinds of hydrogen, in spite of uncertainty about where they will turn up in a ^{1}H NMR spectrum: if the sample, in $CDCl_3$ or CCl_4 solution, is shaken with a drop of D_2O, the OH, NH, and SH hydrogens exchange rapidly with the deuterons, the HDO floats to the surface, out of the region examined by the spectrometer, and the signal of the OH, NH, or SH simply disappears from the spectrum (or, more commonly, is replaced by a weak signal close to δ 4.8 coming from suspended droplets of HDO).

Temperature. The resonance position of most signals is little affected by temperature, although OH, NH, and SH protons resonate at higher field at higher temperatures, because the degree of hydrogen bonding is reduced.

Solvents. Chemical shifts are little affected by changing solvent from CCl_4 to $CDCl_3$ (±0.1 p.p.m.), but change to more polar solvents—such as acetone, methanol, or DMSO—does have a noticeable effect (±0.3 p.p.m.) for ^{13}C and ±0.3 p.p.m. for protons, and benzene can have quite a large effect, ±1 for protons and for ^{13}C. Benzene weakly solvates areas of low electron density; since the benzene has a powerful anisotropic magnetic field (Fig. 3.6b), solute atoms lying to the side of or underneath the solvating benzene ring can experience significant shielding or deshielding relative to their position in an inert solvent like CCl_4. Pyridine is sometimes even more effective. This solvent-induced shift can be very useful when you want to resolve two signals which overlap in the first spectrum that you take.

More substantial shifts can be induced by complexation with paramagnetic salts; this is discussed later in Sec. 3.11.

The common solvents in NMR spectroscopy are used in deuterated form, in order not to introduce extra signals, but they all (except CCl_4) have residual signals from incomplete deuteration. It is important to recognize these signals, listed in Table 3.25, in order to discount them from spectra that you are interpreting. The signal of $CHCl_3$ (at δ 7.25 in ^1H spectra) is evident in many of the spectra taken in $CDCl_3$ used to illustrate this chapter. Fortunately, most of them introduce only one or two sharp signals, and they are easily recognized. The carbon of $CDCl_3$ is very recognizable: it can be seen in Fig. 3.3 as the group of three very weak lines centred at δ 77. The reason that it is three lines and not one takes us into the next section.

3.6   Spin–spin coupling to ^{13}C

We have left out of the discussion so far a significant effect that neighbouring magnetic nuclei have on the signal we detect. If a nearby nucleus has a spin, that spin affects the magnetic environment of the nucleus we are observing, and the signal we detect is not a simple singlet, but a multiplet, the complexity of which is dependent upon the nature and number of the nearby atoms.

^{13}C–^2H coupling
The carbon atom of $CDCl_3$ is attached to a deuterium nucleus. Deuterium has a spin $I = 1$, which means that there are three possible energy levels for a deuterium atom placed in a magnetic field. The carbon atom therefore experiences three slightly different magnetic fields depending upon the spin state of the deuterium nucleus to which it is attached. Since the difference in energy between the three states is very small, there is an essentially equal probability that a carbon atom will be bonded to a deuterium in any one of the three states. The result is that the carbon nucleus comes into resonance at three frequencies with equal probability, as indeed we can see in Fig. 3.3, where the $CDCl_3$ signal is the three equally spaced weak lines at δ 77. The carbon is said to be *coupled* to the deuterium, and the separation of the lines in Hz is called the *coupling constant, J*. Because there is only one bond between the carbon and the deuterium, J is further qualified as $^1J_{CD}$. Carbon–deuterium coupling is much less important than carbon–hydrogen coupling, but we began with it because it is visible in a spectrum we have already used.

^{13}C–^1H coupling
Why is there no carbon–hydrogen coupling in Fig. 3.3? The answer is that while that spectrum was being taken, the sample was irradiated with a strong signal encompassing the whole range of frequencies within which the protons in the molecule came into resonance. This caused the N_α and N_β protons to be exchanging places rapidly several times during the measurement of the carbon signal. Each carbon atom, therefore, 'saw' only an average state for the protons near to it; instead of being coupled, each ^{13}C atom simply gave rise to a single sharp line. ^{13}C spectra are usually taken in this way, and are described as *proton decoupled*.

Let us look at another case, that of the potassium salt of penicillin G. The proton-decoupled spectrum is shown as Fig. 3.8, and it has, as expected, a wide spread of signals over most of the usual range. The two methyl groups are in different stereochemical environments, and appear therefore as the two upfield signals near δ 30. The other five tetrahedral carbons appear between 40 and 80, with the quaternary carbon weaker than the others. The aromatic ring has only four different kinds of carbon atom: the two *ortho* carbons are identical with each other and so are the two *meta* carbons; they will be the two very closely spaced lines of high intensity near 130; the *para* carbon appears at 127, and the *ipso* carbon is the one of low intensity, a little further downfield than the others. Finally, the carbonyl carbons are furthest downfield, and are very weak. They look like only two lines in Fig. 3.8, but expanded they are in fact three. The spectrum, in other words, supports the structure by having the right number of signals for the carbon atoms present, and all in the appropriate

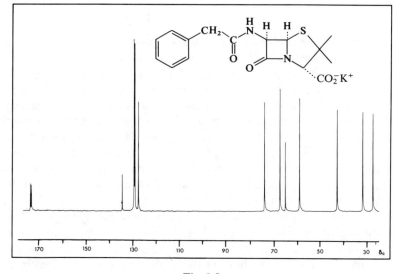

Fig. 3.8

regions. Had the spectrum been taken without proton decoupling, the upfield portion would have looked like that in Fig. 3.9. Here we have the signals of only seven of the carbon atoms, but already it looks formidably complicated.

The easiest signal to find is the quaternary carbon at δ 66: it is weak, as usual for quaternary carbons, and it is still a singlet because it has no hydrogens on it to couple with. The next signals to look for are the five other signals at the low-field end of the spectrum. These are three doublets, A, B, and C, with two of the lines overlapping (at δ 71), making five lines in all. These come from the three methine carbons, in other words the ones which have only one hydrogen atom bonded to them. In each case, the hydrogen atom ($I = \frac{1}{2}$) can take up two orientations with respect to the applied field. The carbon atom therefore 'sees' two different magnetic fields, and comes into

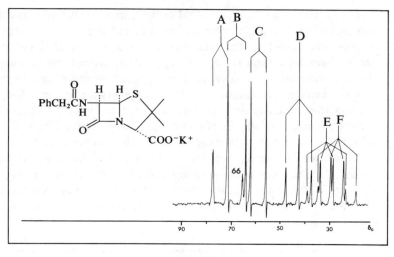

Fig. 3.9

resonance as a doublet (Fig. 3.10a). The two lines of each doublet are equally intense, as you can see in the doublet C, but each doublet A, B, and C is of different intensity from the other two. The next signal upfield, the triplet D, comes from the methylene carbon adjacent to the benzene ring. This carbon is equally coupled to two similar protons. The easiest way to understand the consequence of having two neighbouring protons is to look at the effect of each in turn. The first proton splits the signal into two and the second then splits it again by the same amount, as illustrated in Fig. 3.10b. You can see, from simple geometry, that this creates in the centre a coincident line, made up of two components. The central line is therefore twice as intense as the two outer lines, and the resulting signal is the 1:2:1 triplet, D. Finally, we come to the two upfield signals E and F, each of which is a quartet. These come from the two methyl carbons, each of which is attached to three hydrogens. We can simply extend the argument of Fig. 3.10b to Fig. 3.10c: the first hydrogen splits the carbon into a doublet, the second splits each line of the doublet, creating a triplet as before, and the third splits each of the lines of the triplet into doublets, creating a quartet. Because the three hydrogens are identical, the coupling constants will be identical, and the two central lines will therefore be made up of perfect coincidences. The distribution of intensity within the signal creates a 1:3:3:1 quartet. The eight upfield lines in

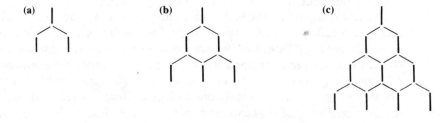

Fig. 3.10

Fig. 3.9 are two overlapping sets of 1:3:3:1 quartets, labelled E and F. In each of the signals shown in Fig. 3.9, the true chemical shift to quote is the centre of the multiplet.

In summary, the signals of quaternary, methine, methylene, and methyl carbons are a singlet, a doublet, a triplet, and a quartet, respectively, Fig. 3.11. These patterns are quite general, and we shall meet them again in proton NMR spectra. The rule, for nuclei of $I = \frac{1}{2}$, is that a nucleus, equally coupled to n others, will give rise to a signal with $n + 1$ lines, and the intensities are given by the coefficients of the terms in the expansion of $(x + 1)^n$.

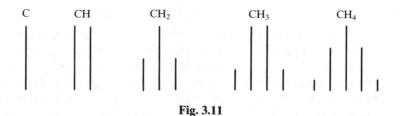

Fig. 3.11

Clearly, the information contained in these multiplets is valuable, but interpretation is complicated by the overlapping of the signals. With larger molecules, it becomes nearly impossible to disentangle the multiplets. The situation is saved by another technique: while the ^{13}C spectrum is being measured, the sample is irradiated at a frequency close to but not coinciding with the resonance frequency of the protons. This is called *off-resonance decoupling*, and has the effect of narrowing the multiplets, without removing them altogether, as in fully decoupled spectra. The upfield portion of the off-resonance decoupled spectrum of potassium penicillinate G is shown in Fig. 3.12, where the doublets, triplets, and quartets are clearly visible and identifiable. Thus it is easily possible to find out how many of each kind of carbon is present in a molecule of unknown structure.

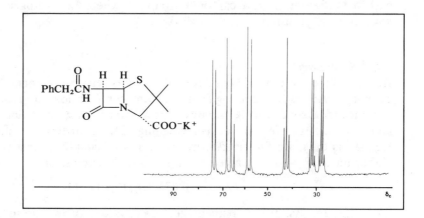

Fig. 3.12

The size of the coupling constant, J, is also informative. It is principally affected by the geometry of the bonds around the carbon atom: tetrahedral (sp^3) carbon usually giving values between 120 and 150 Hz, trigonal (sp^2) carbon between 155 and 205 Hz, and digonal (sp) carbon close to 250 Hz (Table 3.14). The other major influence is the presence of electronegative atoms, which lead to coupling constants at the high end of the range, so that in an extreme case, chloroform has $^1J_{CH} = 209$ Hz. even though it has a tetrahedral carbon. The $^1J_{CH}$ coupling constants of tetrahedral carbon can be estimated using Eq. 3.18 and the data in Table 3.15.

The residual coupling found in off-resonance decoupled spectra is given by:

$$J' = 2\pi J \Delta v / \gamma_H B_2 \tag{3.6}$$

where J is the unperturbed coupling constant, Δv is the difference between the resonance frequency of the coupled proton and the frequency of the decoupling signal, and B_2 is the decoupler power. Thus the coupling constant J is not easily measured directly from off-resonance decoupled spectra, and is more usually measured in the proton NMR spectrum. Most of the proton NMR spectrum is unaffected by the presence of ^{13}C: 99 per cent of the signal from a proton is from those protons attached to ^{12}C, and these are not coupled because ^{12}C is magnetically inert; 1 per cent of the signal, however, comes from protons attached to ^{13}C nuclei, and these are coupled, showing up—when the uncoupled proton signal is a singlet—as a very weak doublet placed symmetrically about the strong signal from the protons attached to ^{12}C, and with the two lines separated from each other by an amount in Hz equal to the coupling constant $^1J_{CH}$. Since each line of this doublet is only 0.5 per cent of the intensity of the main signal, it usually goes unnoticed, but it can be searched for, if that region of the spectrum is not crowded with other signals. When the proton signal is itself a multiplet, the ^{13}C satellites are even weaker, because they are also multiplets.

The coupling of a ^{13}C nucleus to a proton through more than one bond has much smaller coupling constants, only very rarely resolved in off-resonance decoupled spectra, but measurable in proton NMR spectra and in a full ^{13}C spectrum, when the signals are not too confused. $^2J_{CH}$ couplings are at their largest (5–50 Hz) when the carbon is digonal or in a carbonyl group, or when the hydrogen is an aldehyde hydrogen. $^3J_{CH}$ is usually less than 10 Hz.

^{13}C–^{13}C coupling

Because of the low natural abundance of ^{13}C, it is very rare for one ^{13}C to be bonded to another. Any signals coming from such rare combinations are usually too weak to use, but enrichment with ^{13}C is now common in mechanistic and biosynthetic studies, and it is then possible to see the coupling. The geometry (tetrahedral, trigonal, digonal) is the main factor affecting $^1J_{CC}$. The ^1J coupling constants between two carbon nuclei C_x and C_y can be estimated using the expression:

$$^1J_{C_xC_y} = 0.073(\%s_x)(\%s_y) - 17 \tag{3.7}$$

where $\%s_x$ and $\%s_y$ are the percentages of s character (using the sp^n notation) in C_x and C_y. Thus the (tetrahedral) methyl group in toluene is estimated to be coupled to the (trigonal) *ipso* carbon with a coupling constant of 43 Hz; the observed value is 44 Hz. As with $^1J_{CH}$, neighbouring electronegative elements raise the coupling constants; for example, the methyl group of acetates is coupled to the carbonyl carbon with ^1J of 59 Hz, at the upper end of the range of C–C coupling constants.

3.7 ^{1}H–^{1}H first-order coupling

We have already seen with ^{13}C–^{1}H coupling what happens when two nuclei with $I = \frac{1}{2}$ are coupled. Much the same is true for proton–proton coupling, except that we are now concerned with longer-range coupling, principally two-bond and three-bond coupling. $^1J_{HH}$ is only found in hydrogen itself, and is only of importance in the theory of NMR spectroscopy.

15

Let us begin with three-bond coupling, also known as vicinal coupling, found in the arrangement **15**. The coupling constants $^3J_{CH}$ are generally in the range 0–20 Hz. The factors that affect the coupling constant are dicussed later, in Sec. 3.9, but for now we shall look only at the multiplicity. In proton NMR spectra, as with ^{13}C spectra, we get doublets, triplets, and quartets, whenever a proton is coupled equally to one, two, or three protons. For example, Fig. 3.13 shows the ^{1}H NMR spectrum of 1,1,2-trichloroethane.

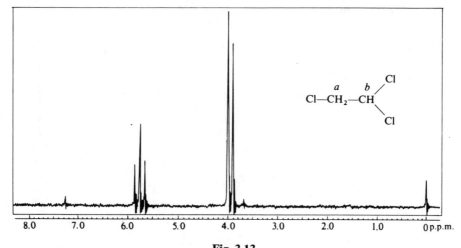

Fig. 3.13

(Varian catalogue, spectrum no. 2.)

The downfield signal, centred at δ 5.77, is the signal from the methine hydrogen b, downfield because it has two electronegative elements attached to the methine carbon. It resonates as a 1:2:1 triplet, because the methine hydrogen is coupled to the two identical methylene hydrogens a. Likewise, the upfield signal at 3.68 comes from the methylene hydrogens a and has an appropriate chemical shift for a hydrogen on a carbon carrying only one electronegative element (see Table 3.17). It appears as a doublet because the two hydrogens of the methylene group are coupled to the single methine hydrogen, and are split into two by it. The two methylene hydrogens are, of course, identical; they experience identical magnetic environments, and they come into resonance at exactly the same place. The upfield doublet, therefore, is twice as intense as the downfield triplet. The methylene protons are, in fact, coupled to each other, but *coupling between protons with identical chemical shifts does not show up in NMR spectra*. In summary, the downfield signal is a one-proton *triplet* because the proton which gives rise to it is equally coupled to *two* protons, and the upfield signal is a two-proton *doublet* because the protons which give rise to it are coupled to *one* proton. The rule is the same as that given in the section on $^{13}C^{-1}H$ coupling: a nucleus, equally coupled to n others, will give rise to a signal with $(n+1)$ lines, and the intensities are given by the coefficients of the terms in the expansion of $(x+1)^n$. It is worth remembering how small the outer lines get in such an expansion: the outer lines of a septet, for example are very weak, and the signal is easily mistaken for a quintet.

Another example is provided by the spectrum of ethanol (Fig. 3.14). The upfield signal is a three-proton triplet coming from the methyl group a. The chemical shift of this signal (δ 1.22) is appropriate (Table 3.17) for a C-methyl group with an electronegative element nearby. The three hydrogens of the methyl group do not couple visibly with each other, because they are identical, but they couple equally

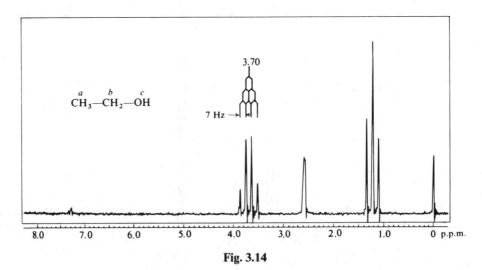

Fig. 3.14

(Varian catalogue, spectrum no. 14.)

with the two protons of the methylene group. Likewise the two protons of the methylene group *b* adjacent to the OH group are identical and come into resonance at appropriately lower field (δ 3.70). The signal is a 1:3:3:1 quartet, because the methylene protons are equally coupled to the three hydrogens of the methyl group. This pattern of a three-proton triplet and two-proton quartet is characteristic of an ethyl group in which the methylene protons are not coupled to anything else other than the methyl group.

The coupling of the methylene hydrogens to the proton *c* of the OH group does not appear, either in the methylene signal or in the OH signal at δ 2.58. This coupling can appear when the sample is exceptionally pure, or when the spectrum is taken in d_6-DMSO. When it does, the methylene group appears as a doubled quartet, in other words a 1:4:6:4:1 quintet, and the OH signal appears as a triplet. It does not appear in the spectrum shown in Fig. 3.14, because when that spectrum was taken the OH groups were rapidly exchanging their protons intermolecularly, almost certainly in a reaction catalysed by traces of acid or base in the sample or in the tube. If the rate of the exchange is appreciably faster than the difference in frequency between the lines of the triplet (6 Hz in this case), the receiver detects the hydroxyl proton in an average of the three magnetic microenvironments created by the three possible arrangements of the nuclear magnets of the methylene protons. In very pure samples and in d_6-DMSO, the rate constant for the exchange is less than 6 s^{-1}, and the coupling is then visible. Coupling involving NH protons in amines and SH protons is similarly visible only in special cases. The signals of the OH, NH and SH groups are removable (see page 75) with a D_2O shake. However, CH groups adjacent to amide NH groups usually show coupling ($J=5$–9 Hz), even when the amide NH signal is very broad from quadrupole relaxation.

A slightly more complicated example is shown in the spectrum of 1-nitropropane (Fig. 3.15). The methyl group *a* is again a three-proton triplet at high field (δ 1.03),

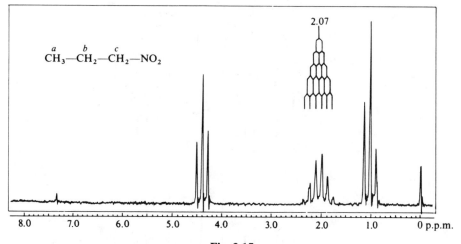

Fig. 3.15

(Varian catalogue, spectrum no. 42.)

because it is adjacent to a methylene group. The methylene group c is a two-proton triplet, at low field (δ 4.38). The chemical shift is appropriate for a methylene group next to an electronegative and anisotropic group, and the multiplicity is appropriate for protons coupling to a methylene group. The methylene group in the middle b is a two-proton $1:5:10:10:5:1$ sextet at 2.07. The chemical shift is appropriate for a methylene group between two alkyl groups, but not far from an electronegative group. The multiplicity is appropriate for protons coupling equally to a total of five protons. In this case, the coupling constant J_{ab} is almost exactly the same as J_{bc}, but the slight broadening of the lines in the sextet is a consequence of the equality being not quite perfect.

In general, if a proton has as neighbours sets n_a, n_b, n_c, ... of chemically equivalent protons, the multiplicity of its resonance will be $(n_a + 1)(n_b + 1)(n_c + 1)$.... This gives rise to a great many possible multiplet patterns, and it is important to learn how to recognize and analyse them. Thus, the methylene group b could have appeared as a triple quartet, in other words 12 lines, but did not because of the accidental coincidence of the coupling constants. The analysis in Fig. 3.16 is one possible appearance for a triple quartet, and illustrates the way in which such signals can be analysed and hence recognized. This particular case has J_{AB} and J_{BC} in a ratio of $5:3$, but, whatever the ratio, the intensity pattern is $1:2:3:1:6:3:3:6:1:3:2:1$, made up by combining $1:2:1$ and $1:3:3:1$ fragments.

In all three spectra so far, the weak signal at δ 7.25 is from the chloroform present in the deuterochloroform solvent. In all three spectra, the coupling constants have been very much the same, 6–7 Hz, typical of coupling constants in freely rotating alkyl chains. Finally, in all three spectra, the separation of the signals (in Hz) has

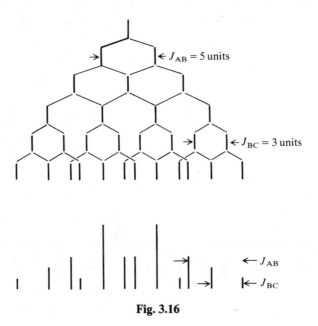

Fig. 3.16

been much greater than the coupling constants (in Hz), and this has allowed us to interpret the spectrum using what is called the *first-order approximation*.

16

Two-bond coupling, $^2J_{H_aH_b}$, also known as geminal coupling, is found only in methylene groups **16** in which for some reason the two hydrogens *a* and *b* are not identical, and do not therefore come into resonance at the same frequency. We shall see examples of this later, but for now it is only necessary to establish that they give rise to multiplets, in the first-order approximation, with the same rules as for three-bond coupling, and the range of coupling constants is rather similar, 0–25 Hz. However, two hydrogens bonded to the same carbon atom are frequently close in chemical shift, and first-order analysis is sometimes not possible. We must now look into some of the consequences, both for two-bond and for three-bond coupling, of having the chemical shift difference and the coupling constant more alike.

3.8   Some simple ^1H–^1H splitting patterns

The convention used is to label protons close in chemical shift with the letters A, B, and C, those far away in chemical shift from these with the letters X, Y, Z, and those intermediate with the letters M, N, and O. Thus, we have already seen an AX_2 pattern in Fig. 3.13, an A_2X_3 pattern in Fig. 3.14, and an $A_3M_2X_2$ pattern in Fig. 3.15.

AB Systems
An AB system consists of two mutually coupled protons A and B, which are not coupled to any other protons, and which have the difference in chemical shift between the A and the B signal $\delta_{AB} = \delta_A - \delta_B$ comparable in magnitude to the coupling constant J_{AB}. The comparison is with the simple, first-order spectrum of an AX system, where we can expect the A and the X signals to be a pair of well separated doublets (Fig. 3.17a), with each line equal in height.

The AB patterns are shown in Fig. 3.17b and 3.17c. The coupling constants stay the same:

$$J_{AB} = v_4 - v_3 = v_2 - v_1 \qquad (3.8)$$

but, as the chemical shift difference gets less, the inside lines, 3 and 2, grow in size, and the outside lines, 4 and 1, get smaller. The closer the signals are in chemical shift, the more the distortion (Fig. 3.17c). In an AX system, the chemical shifts of the A and the X signals are given by the frequency of the midpoints of each of the doublets. This is no longer the case with an AB system, where the chemical shift is given by:

$$\delta_A - \delta_B = \sqrt{(v_4 - v_1)(v_3 - v_2)} \qquad (3.9)$$

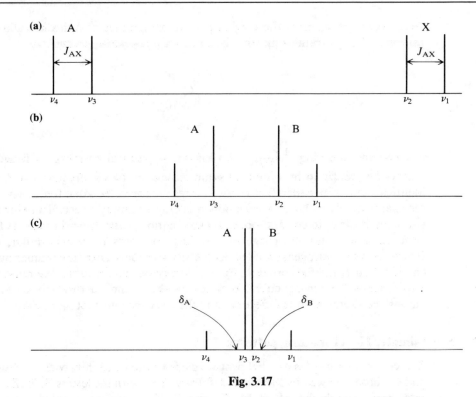

Fig. 3.17

As illustrated in Fig. 3.17c, this places the true chemical shifts closer to the inside lines than to the outside lines. The relative intensity of the lines, I, is given by:

$$I_3/I_4 = I_2/I_1 = (v_4 - v_1)/(v_3 - v_2) \tag{3.10}$$

These patterns are very helpful in identifying AB systems: a strongly distorted pair of lines must be coupling to a proton close in chemical shift, and a less distorted doublet to one further away. Furthermore, the distortion tells one in which direction to look for the other half of the AB system. The doublet is described as *pointing* to its partner. At the extreme, when A and B have exactly the same chemical shift, the outside lines disappear, and the inside lines merge into a singlet. This is the situation with the methylene groups in the three compounds used in Figs 3.13, 3.14, and 3.15. In these three cases, the methylene hydrogens are inherently identical, but coupling also disappears when two chemically dissimilar protons accidentally come into resonance at the same frequency.

The ^{1}H NMR spectrum of columbianetin (**17**), shown in Fig. 3.18, has two AB systems. The one-proton doublet centred at 7.65 is associated with the proton i, which comes into resonance at low field because it is β to a carbonyl group (Table 3.3). The corresponding α proton, f comes into resonance at 6.22, substantially further upfield. The two signals show the typical AB pattern for protons fairly well separated in chemical shift. The actual chemical shifts of the four lines in this 60 MHz spectrum

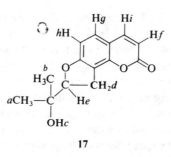

17

are 6.13, 6.29, 7.58, and 7.74 p.p.m. In frequency units these are 368, 377, 455, and 464 Hz, from which we can calculate that δ/J ($\sim$9) is not particularly small, and the line perturbation (Eq. 3.9) is not very great: $I_3/I_4 = I_2/I_1 = 1.23$.

The second AB system is produced by the protons h and g, which come into resonance at 6.79 and 7.29. In this case, $J_{hg} = 8$ Hz and $\delta_{hg} = 30$ Hz, and δ/J is smaller ($\sim$4). Consequently, the line perturbation is greater, as the spectrum shows. It is easy in this case to pair up the appropriate AB systems. The degree of distortion in each matches rather well, and the coupling constants are different: the pair i and f have a coupling constant of 9.5 Hz, and the pair h and g have a coupling constant of 8 Hz. It is always wise when you are assigning signals to measure up doublets, in order to make sure that both halves of what you think are an AB system have matching coupling constants. This is equally true for more complicated systems. The remaining signals of the spectrum can easily be assigned. The one-proton signal at 2.00 is from the OH group c. The methylene group, which has two chemically different hydrogens d, one *cis* to H$_e$ and one *trans* to it, nevertheless accidentally come into resonance at the same frequency, and do not show any coupling with each other; they appear as

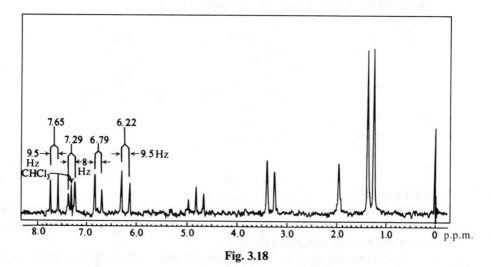

Fig. 3.18

(Varian catalogue, spectrum no. 310.)

the doublet at 3.33. The methine hydrogen *e* is coupled with equal coupling constants to both of the methylene hydrogens *d*, and appears therefore as an approximately 1:2:1 triplet at 4.82. Notice again that the separation of the lines of the triplet and the doublet match, as they must if they are mutually coupled, and notice also that these multiplets point at each other to a small but noticeable extent, the upfield line of the triplet being larger than the downfield line.

The two singlets at 1.25 and 1.37 are from the two methyl groups *a* and *b*. The two lines are not described as a doublet, because they are not a result of coupling. At first sight, you might expect the methyl groups to be chemically identical, especially when you allow for the free rotation of the side chain. It is important to understand why they are different, because this phenomenon is common in NMR spectra, both for pairs of methyl groups adjacent to a chiral centre, as here, and even more common for two hydrogens in a methylene group adjacent to a chiral centre (see Fig. 3.20). It is even quite commonly observed when the chiral centre is further away. In the first place, the side chain will almost certainly adopt one conformation in preference to any other, and in that conformation the two methyl groups will not be in the same environment. Secondly, even if the rotation were completely free, the average field experienced by the one methyl group, say *a*, as it moved around, would not inherently be the same as that experienced by the other, *b*. At any one monent, frozen as **18**, when methyl group *a* is at the top, arbitrarily defined, it will have the methyl group *b* and the OH group next to it in a clockwise order. When methyl group *b* comes to the top, **19**, it will have the methyl group *a* and the OH group in an anticlockwise order. The two conformations **18** and **19** are not identical, nor are they enantiomers. In other words, the methyl groups do not sweep out the same environment as the side chain rotates. The two methyl groups are said to be *diastereotopic*. Only when the average field is the same by coincidence do diastereotopic methyl groups, and diastereotopic hydrogens likewise, come into resonance at the same frequency, as in fact is the case with the methylene group *d* in this molecule.

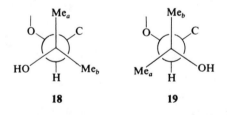

ABX systems

ABX patterns have a wide variety of appearances, depending upon the relative chemical shifts and coupling constants of the three protons. The example given is only one possible arrangement.

If we combine the treatment of an AB system from the previous section with a simple first-order prediction from Sec. 3.7, we can expect that the four AB lines will each be split into doublets by the X proton, while the X proton—being chemically well shifted from A and B, but coupled to both—should appear as a double doublet. The AB lines will inherit the differences in intensity stemming from their AB coupling, but the double doublet of the X proton should have four more or less equally intense lines. The prediction then is a pattern like that illustrated in Fig. 3.19, with a total of 12 lines, but there are many possible versions, depending upon whether any of the lines happen to coincide, and depending upon how large the separation of the A and B signals is. Thus, if A and B are closer in chemical shift, line 5 could easily come at lower field than line 4 or line 3. This type of pattern is quite commonly found, but it is wise to be careful in assigning coupling constants to real systems using this simple analysis. The separation of the lines 9–12 only gives a rough indication of the coupling constants: J_{AX} is close to the separation of lines 9 and 10, and J_{BX} is close to the separation of lines 9 and 11, but it is strictly true only that the separation of lines 9 and 12 is the sum of J_{AX} and J_{BX}.

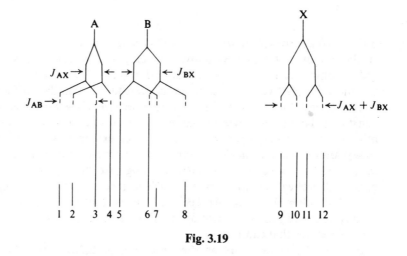

Fig. 3.19

Patterns like this, and more complicated versions of the same idea, turn up frequently in NMR spectra. For example, Fig. 3.20 shows the signals of the three carbon-bound protons of sodium aspartate. The geminal pair, H_β and $H_{\beta'}$, are diastereotopic (Sec. 3.8); they are coupled to each other with the largest of the three coupling constants (16 Hz), and they are each coupled to H_α with different coupling constants, one large (10 Hz) and one small (4 Hz). The major difference from the pattern in Fig. 3.19 is that the X signal in this example, H_α, is downfield from the A and B signals, rather than upfield from them. Notice how the two lines from $H_{\beta'}$ at highest field (δ 2.6) are nearly equal in intensity, as are the downfield pair of lines, showing that these pairs of lines identify coupling to a signal far away in chemical shift, H_α (δ 4.5), whereas the upfield pair are significantly smaller than the downfield pair, showing that this coupling is to a proton close in chemical shift, H_β (δ 3).

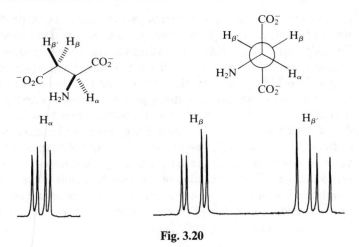

Fig. 3.20

(Reproduced with permission from J. K. M. Sanders and B. K. Hunter, *Modern NMR Spectroscopy*, OUP, Oxford, 1987.)

Substantial deviation from the first-order approximation
In the AB and ABX systems above, we have not, in fact, strayed far from the first-order approximation; we have only added the distortions that result when the chemical shift difference and the coupling constant are similar, and added a cautionary word that the position of the lines may not strictly allow you to measure the coupling constants. There are, however, more substantial failures of the first-order approximation when several spins are involved. For example, a pair of mutually coupled multiplets close in chemical shift can have the upfield branch of one and the downfield branch of the other overlap, and the resulting pattern is not always first order. With a proper theoretical treatment, it is possible to account for these patterns, and for those described below. For now, we need only know that some splitting patterns are not readily analysed just by looking at them. Three examples will suffice to illustrate the kind of pattern that can turn up.

Figure 3.21 shows part of the spectrum of 2,2-dimethyl-4-cyanobutanal (the aldehyde proton is at δ 9.47). The adjacent methylene groups might have been expected merely to couple with each other to give a pair of triplets, as adjacent methylene groups frequently do. The pattern produced in this case is, however, clearly more complicated. Similarly, in 1,4-diphenylbutadiene, with only two kinds of chemically different olefinic proton, we might expect an AB system in the olefinic region. In fact, the olefinic region (Fig. 3.22) is substantially more complicated, although it does bear a superficial resemblance to an AB system. Finally, the characteristic appearance of the aromatic proton resonances for a *para* disubstituted benzene ring is illustrated in the spectrum of *p*-bromophenetole (Fig. 3.23), where the two different kinds of proton are coupled together in a pattern that again looks superficially like an AB system, but actually has a number of extra lines, of which the 'inside lines' are the most noticeable. The complexity in these three cases comes from the fact that the pairs of *a* nuclei (and *b* nuclei) are not magnetically equivalent. For

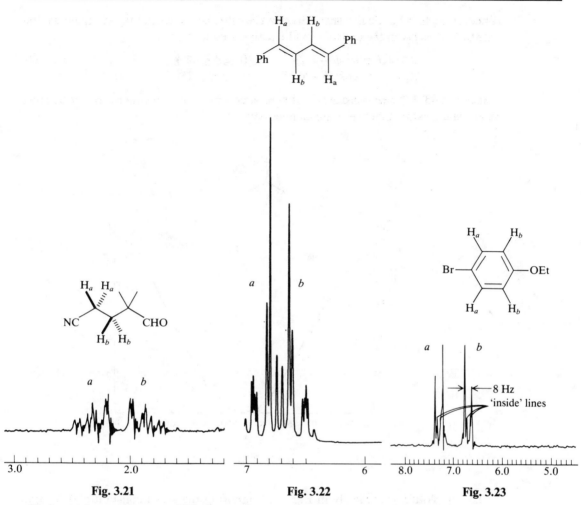

Fig. 3.21   **Fig. 3.22**   **Fig. 3.23**

(Varian catalogue, spectrum no. 178.)

example, in *p*-bromophenetole, a given *a* nucleus has very different coupling constants to the two *b* nuclei. All three cases are examples of AA′BB′ systems.

3.9   The magnitude of ^{1}H–^{1}H coupling constants

In the course of the discussion so far, we have seen *J* values as widely different as 1 Hz and 16 Hz. We must now consider what factors most significantly affect the size of *J*.

Vicinal Coupling (^3J, *H*—*C*—*C*—*H*)

The dihedral angle.   Coupling is mediated by the interaction of orbitals within the bonding framework. It is therefore dependent upon overlap, and hence upon the dihedral angle. The relationship between the dihedral angle and the vicinal coupling constant ^3J, is given *theoretically* by the Karplus equations:

$$^3J_{ab} = J^0 \cos^2 \phi - 0.28 \qquad (0° \leq \phi \leq 90°)$$
$$^3J_{ab} = J^{180} \cos^2 \phi - 0.28 \qquad (90° \leq \phi \leq 180°)$$

(3.11)

where J^0 and J^{180} are constants which depend upon the substituents on the carbon atoms and ϕ is the dihedral angle defined by:

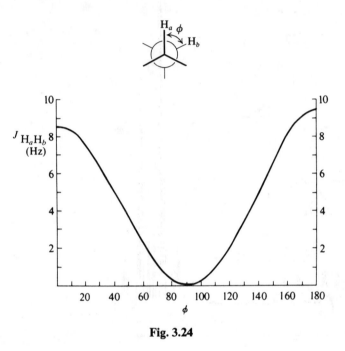

Fig. 3.24

This is plotted graphically in Fig. 3.24, ignoring the small constant of 0.28, and using $J^0 = 8.5$ and $J^{180} = 9.5$, the standard values when no better estimate is available. Coupling constants observed experimentally follow this relationship well, but it is not always easy to choose values of J^0 and J^{180}. The coupling constant is at its largest when the dihedral angle is 180°, in other words, when the hydrogens are antiperiplanar and the orbitals are overlapping most efficiently; slightly smaller when it is 0°, when they are syncoplanar; and at its lowest when the dihedral angle is 90°, and the orbitals are orthogonal. In an ethyl group, the free rotation allows the vicinal hydrogens to pass through all these angles, but they will spend most of their time in the usual staggered conformation, with dihedral angles of 60°, 120°, and 180°. The coupling constants of 6–7 Hz that we have seen in Figs 3.12 and 3.13 are approximately the average of the coupling constants for these three angles. In rigid systems, where averaging is not possible, we frequently get both larger and smaller values. In rigid cyclohexanes **20**, for example, the axial–axial coupling constant, J_{aa}, is usually large, in the range 9–13 Hz, because the dihedral angle is close to 180°. The axial–equatorial

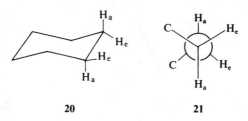

20   21

and equatorial–equatorial coupling constants, J_{ae} and J_{ee}, are much smaller, usually in the range 2–5 Hz, because the dihedral angles are close to 60°. The dihedral angles are clearer on the Newman projection **21**, but it should be remembered that the bond angles are not quite so perfectly bisected in real systems.

We can now see why the vicinal coupling constants in the spectrum in Fig. 3.20 are so different. The dianion of aspartate will adopt the conformation shown in the Newman projection in Fig. 3.20, with the two anionic groups as far apart as possible. This gives H_β and H_α a dihedral angle of 60° and hence a small coupling constant, and $H_{\beta'}$ and H_α a dihedral angle of 180° and a large coupling constant.

A modified Karplus equation can be applied to vicinal coupling in alkenes; the numbers are slightly different, but the conclusion is the same. A dihedral angle of 180° is found in *trans* double bonds **22**, and the coupling constants are large, and a dihedral angle of 0° is found in *cis* double bonds **23**, and the coupling constants are smaller.

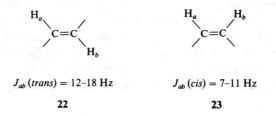

J_{ab} (*trans*) = 12–18 Hz J_{ab} (*cis*) = 7–11 Hz

22   23

The presence of electronegative or electropositive elements.   Electronegative elements directly attached to the same carbon atom as one of the vicinally coupled protons reduce the coupling constant, and electropositive elements raise it. For freely rotating chains, the effect is small (**24–26**).

ClCH$_2$CHCl$_2$	CH$_3$CH$_2$Cl	CH$_3$CH$_2$Li
$^3J = 6.0$ Hz	$^3J = 7.3$ Hz	$^3J = 8.4$ Hz
24	25	26

When an electronegative element is held rigidly antiperiplanar with respect to one of the protons (heavy outline in **27**), then the effect is larger. Thus J_{ae} is only 2.5 ± 1 Hz when X in **27** is OH, OAc, or Br, but it is 5.5 ± 1 in **28**, even though the dihedral angles are close to 60° in both cases. On double bonds, both the antiperiplanar and the syncoplanar protons are affected. An electronegative element lowers both the *cis* and the *trans* coupling constants (**29**, X = F), and an electropostive element has the

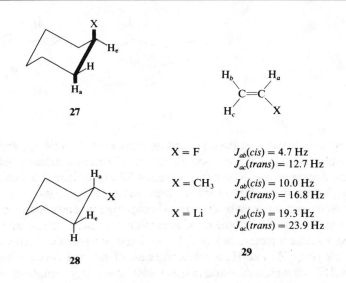

27

28

29

X = F	$J_{ab}(cis) = 4.7$ Hz
	$J_{ac}(trans) = 12.7$ Hz
X = CH$_3$	$J_{ab}(cis) = 10.0$ Hz
	$J_{ac}(trans) = 16.8$ Hz
X = Li	$J_{ab}(cis) = 19.3$ Hz
	$J_{ac}(trans) = 23.9$ Hz

opposite effect: the *cis* coupling constant for vinyl-lithium (**29**, X = Li) is higher even than the normal value for a *trans* double bond, and the *trans* coupling is higher still.

We can see an extreme example of the lowering of the coupling constant by an electronegative element in the spectrum in Fig. 3.25. 3,4-Epoxytetrahydrofuran has a plane of symmetry, and therefore shows only three resonances: an AB system from the methylene protons *a* and *c*, and a sharp singlet at δ 3.79 from the methine proton *b*. The dihedral angle between *b* and *c* is close to 90°, and it is not suprising that there is no coupling between them. However, the dihedral angle between *b* and *a* is somewhere between 0° and 30°, and a coupling constant of 6–8 Hz would be expected on the basis of the Karplus equation. A major influence is the epoxide oxygen *anti* to proton *a*, but angle strain also contributes to the disappearance of coupling.

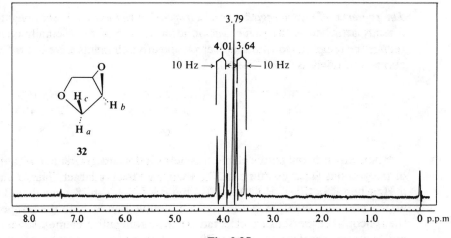

Fig. 3.25

Angle strain. In the fragment **30**, ^3J decreases as θ and θ' increase. This effect is noticeable in the *cis* coupling constants between the olefinic protons in cyclic alkenes **31**: as the ring size changes, the coupling constant changes. It is therefore possible to tell in many cases into what size ring a double bond is incorporated.

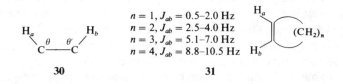

$n = 1, J_{ab} = 0.5\text{–}2.0 \text{ Hz}$
$n = 2, J_{ab} = 2.5\text{–}4.0 \text{ Hz}$
$n = 3, J_{ab} = 5.1\text{–}7.0 \text{ Hz}$
$n = 4, J_{ab} = 8.8\text{–}10.5 \text{ Hz}$

30   **31**

Bond-length dependence. Aromatic carbon–carbon bonds have bond lengths intermediate between normal single and double bonds. In consequence, *ortho* coupling constants are typically about 8 Hz, rather lower than *cis* olefinic coupling constants in cyclohexenes (8.8–10.5 Hz).

Some of the influences on coupling constants described above can be seen in the spectrum (Fig. 3.26) of **32**. The three-proton singlet at 2.45 and the two-proton singlet at 6.95 correspond to protons *a* and *d*, respectively. Protons *b* and *c* give rise to an AB system; the signals are centred at 6.88 and 7.30 (not necessarily respectively) and the

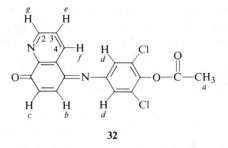

32

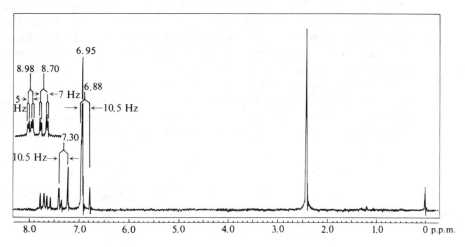

Fig. 3.26

large *cis* olefinic coupling constant (10.5 Hz) indicates that the double bond is not in an aromatic ring. Proton *e* resonates as a double doublet ($J_{2,3} = 5$ Hz and $J_{3,4} = 7$ Hz) centred at 7.68. These couplings are matched in the signals from the protons *g* and *f*: the former at lowest field, 8.98, because it is adjacent to the nitrogen atom of the pyridine ring, and the latter only a little upfield, 8.70, because is still experiences deshielding as a result of the withdrawal of electrons from the 4-position of the pyridine ring. Thus, we can assign the coupling of 5 Hz to J_{eg} and the coupling of 7 Hz to J_{ef}. Both of these signals show a further splitting of 2 Hz, because protons *g* and *f* are coupled to each other in a longer range coupling, ^4J.

Tables 3.27 and 3.28, which you will find with other data at the end of this chapter, summarize vicinal coupling constants.

Geminal coupling (^2J, H—C—H)
In the mathematical treatment of coupling, the three-bond couplings discussed above, and the one-bond ^{13}C–^{1}H couplings discussed earlier, are positive in sign. Two-bond coupling, however, is negative in sign. This has no effect upon the appearance of the spectrum, but it does change the way structural variations affect the coupling constant. This type of coupling only appears in a spectrum when the two protons attached to the same carbon atom come into resonance at different frequencies. However, the coupling constants can be measured, even in molecules such as methane, by introducing a deuterium atom, and measuring the geminal coupling from H to D. The value obtained is related to the proton–proton coupling constant by:

$$J_{\text{HH}} = 6.55 \, J_{\text{HD}}$$

which applies to all coupling, whether geminal or not.

Adjacent π-bonds. The ^2J coupling constant for a simple hydrocarbon, such as methane, is -12 Hz, **33**. When the C—H bond is able to overlap with a neighbouring π-bond as in **34**, the coupling constant is made more negative, in other words effectively larger. The effect is greater when the hyperconjugation is with the π-bond of a carbonyl group than when it is simply with a C=C double bond. Thus in acetone the coupling constant is -14.9 Hz, and in toluene it is -14.3 Hz. The methyl group in acetone is freely rotating, and the measured coupling constant of 14.9 Hz is an average of the coupling between the geminal hydrogens for all the conformational relationships they find themselves in. In rigid and especially cyclic systems in which the conformation is held favourably for overlap such as that in **34**, it commonly reaches -16 or -18 Hz, and if the hyperconjugation is to two double bonds flanking a methylene group, the coupling constant can reach -20 Hz or more.

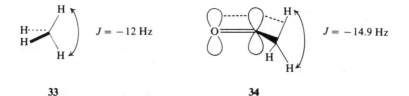

33   34

Adjacent electronegative elements. In contrast to a π-bond, which is effectively electron withdrawing, an electronegative element adjacent to the methylene group is effectively a π-donor with respect to the C—H bond, donating electrons into the antibonding σ^*-orbital, **35**. The coupling constant is now more positive, in other words smaller.

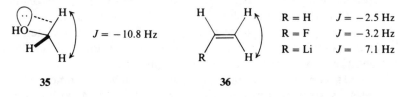

35   **36**

Angle strain. An increase in the H—C—H angle makes ^2J more positive, in other words smaller. This effect is most noticeable in the methylene groups of alkenes, **36**, where the angle has reached something near to 120°, and the coupling constant is close to zero. This coupling is very dependent upon the nature of substituents at the other end of the π-bond, electronegative elements, e.g. fluorine, making them more negative, and electropositive elements, e.g. lithium, actually making the coupling positive and quite large. The effect of the H—C—H angle is also seen in the ranges of ^2J for cycloalkanes (Table 3.26).

The spectrum of 2-methyl-4-oxacyclopentanone (Fig. 3.27) shows a geminal coupling between protons d and e. These protons are not coupled to anything else, and they appear, therefore, as a simple AB system at 3.84 and 4.07. The large coupling constant (-16.5 Hz) is consistent with their being adjacent to and overlapping efficiently with a π-bond. The protons c and f are also coupled to each other, but in addition they are coupled to proton b. As it happens, all three coupling constants are nearly equal in absolute magnitude, $^2J_{cf} = {}^3J_{bc} = {}^3J_{bf} = |8.5 \text{ Hz}|$, so that both c and f

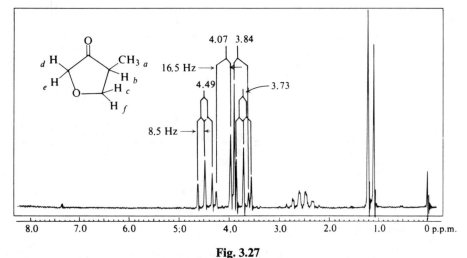

Fig. 3.27

(Varian catalogue, spectrum no. 438.)

appear as triplets centred at 3.73 and 4.49. The low geminal coupling constant is reasonable for a methylene group adjacent to and overlapping efficiently with an atom bearing a lone pair. The equality of the *cis* and *trans* vicinal coupling constants is not uncommon in five-membered rings; such coupling constants cannot safely be used to assign stereochemistry in the way that they can in six-membered rings. The proton *b* will also be a triplet from its coupling with protons *c* and *f*, but it is further coupled to the methyl protons, *a*, with a coupling constant that is close to 8.5 Hz. The resulting signal at 2.55 is approximately a 1:5:10:10:5:1 sextet, but the fine splittings evident in this signal reveal that the coupling constants are not all exactly 8.5 Hz. The methyl group is responsible for the doublet at 1.17.

The spectrum of the isomer 3-methyl-4-oxacyclopentanone (Fig. 3.28) again shows the AB system from protons *d* and *e* at 3.83 and 4.04, with a large coupling constant (-16.5 Hz). The protons *b* and *c*, resonating at 2.15 and 2.57, are also adjacent to a carbonyl group, and couple with each other with an even larger coupling constant (-18 Hz), larger because they are not adjacent to the oxygen atom, as protons *d* and *e* are. The other coupling constants, J_{bf} and J_{cf}, are now different (5.5 and 9 Hz, respectively), making the signals of protons *b* and *c* into double doublets; in other words, the usual eight lines of the AB part of an ABX system. Proton *f*, coupled to the methyl protons *a* with J_{af} of 6 Hz, experiences nearly equal coupling constants (~ 6 Hz) with a total of four protons, and one different coupling constant (9 Hz) with one proton. The result is a 6 Hz 1:4:6:4:1 quintet doubled by 9 Hz, a 10-line pattern, which can be seen at δ 4.35 (except that the eighth line is hidden under the most downfield line of the AB system from protons *d* and *e*).

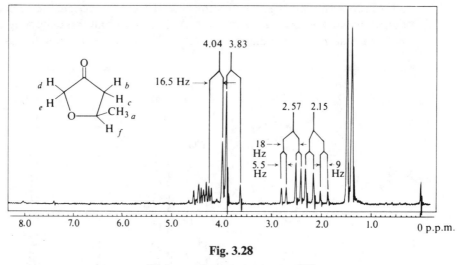

Fig. 3.28

(Varian catalogue, spectrum no. 439.)

Long-range coupling

Allylic and homoallylic coupling.   Coupling through four or more bonds is often called long-range coupling. The coupling constants are naturally quite small, rarely outside the range 0–3 Hz. They are at the high end of the range ($^4J = 2$–3 Hz) when a double bond is one of the four intervening bonds. Allylic coupling is most likely to be resolved ($^4J = 2$–3 Hz) when the allylic C—H bond can overlap with the double bond, as in the allene, propargyl, and allyl partial structures **37**. The difference between the *cisoid* and *transoid* coupling constants are occasionally useful in assigning geometry, but are not very reliable. Homoallylic coupling is also often resolved ($^5J = 1$–2 Hz), but only when the allylic C—H bonds overlap with the double bond, as in the allene and allyl partial structures **38**. When the overlap of the orbitals is especially favourable, long-range coupling can give rise to remarkably high coupling constants, as in a 1,4-cyclohexadiene (**39**).

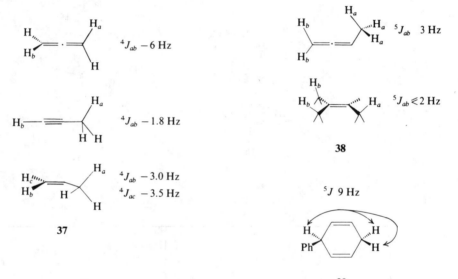

W coupling. In saturated systems, four-bond coupling is again mediated by orbital overlap, which is at its most effective when the four bonds adopt a W arrangement, as emphasized for the 1,3-diequatorial protons in rigid cyclohexanes **40**. Again, there are exceptionally high values when the overlap of the σ-bonds is especially favourable, as in **41**. W coupling is also evident in unsaturated systems, as in the frequently resolved *meta* coupling in benzene rings **42**. The five-bond coupling between *para* protons is not usually resolved.

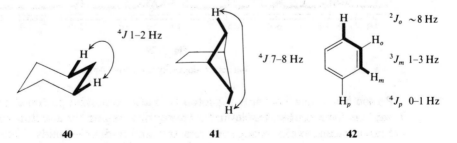

The spectrum (Fig. 3.29) of plumericin **43** shows typical allylic coupling, and will allow us to go through the features of a moderately complicated spectrum, by way of summary. Note, as we go through these assignments, that the chemical shifts are in the appropriate regions. Use the tables of chemical shifts at the end of this chapter for guidance. Assigning first the more obvious signals, the singlet at 3.79 is from the methoxyl group, and the doublet at 2.11 ($J_{ai} = 7$ Hz) is the methyl group a of the ethylidene group. Proton f is adjacent to two oxygen atoms, and should therefore be at low field: it is the doublet centred at 5.59. The coupling constant, J_{bf}, of 6 Hz is consistent with a small dihedral angle somewhat larger than 0° (see **44**).

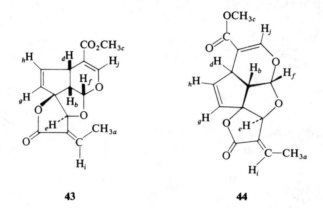

43   **44**

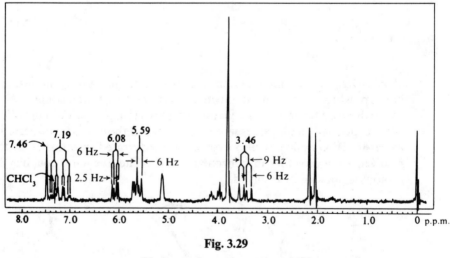

Fig. 3.29

(Varian catalogue, spectrum no. 640.)

Proton b is coupled not only to proton f but also to another proton d. To find proton b, we look for a double doublet in an appropriate region for a methine proton close, but not adjacent to electronegative elements and to double bonds. Thus, we find the signal of proton b at 3.46: it has the coupling of 6 Hz, matching that of proton f, and it has a larger coupling of 9 Hz, appropriate for a dihedral angle between protons b and d that is very close to 0°. We may note also that the 9 Hz coupling is 'pointing' strongly downfield, but the smaller coupling is not pointing significantly, indicating that it is indeed coupling to a proton resonating at a very different frequency. The signal at 5.59 from proton f is misleading: it appears to be pointing downfield. This is because the low-field line of the doublet is superimposed upon another signal, and is thereby given a spurious intensity.

We now look for the signal with a matching coupling of 9 Hz, and we know to look slightly downfield, not only because we have already seen a signal strongly 'pointing' in that direction, but also because the proton d is doubly allylic, and can be expected to come into resonance at δ 3.5–4.5. We find the signal at 4.0 as a double triplet, with the doublet component ($J = 9$ Hz) pointing strongly upfield. The proton d is evidently coupled with protons h and g to approximately equal extents (2.5 Hz), in spite of the fact that one is a three-bond coupling and the other a four-bond (allylic) coupling. Clearly, the dihedral angle between protons d and h is approaching 90°, which minimizes the three-bond coupling and maximizes the allylic coupling to proton g. Protons g and h themselves are in the olefinic region at 5.67 and 6.08 (not necessarily respectively); the latter is clearly resolved as a double doublet with a J_{gh} of 6 Hz, appropriate for an alkene in a five-membered ring. In contrast the proton j, the most downfield signal in the spectrum, because of the deshielding effect of the *cis* methoxycarbonyl group, is a sharp singlet at 7.46. Evidently the allylic coupling from it to proton d is too small to be resolved, presumably because the C—H$_d$ bond is held at an angle that prevents effective overlap with the double bond in the six-membered ring.

The remaining allylic system again shows allylic coupling: the olefinic proton i is a well-resolved double quartet at 7.19, with a large coupling to the methyl group (7 Hz) and a small coupling (1.5 Hz) to the allylic proton e. The proton e itself is not well resolved: it is responsible for the broad singlet at 5.1. The signal is almost certainly broadened by unresolved long-range coupling, a common problem in rigid polycyclic structures.

3.10   Line broadening and environmental exchange

Even in the absence of spin–spin coupling, some lines in NMR spectra are broad. There are two common reasons for this: efficient relaxation and environmental exchange.

Efficient relaxation

Relaxation is promoted by local variations in the magnetic field. We have already seen (Sec. 3.4) how the unpaired spin of paramagnetic salts causes more rapid relaxation, increasing the intensity of ^{13}C signals from those carbon atoms relaxing at unusually slow rates. However, if the atom is already relaxing fast, as with most protons, the effect of an even faster relaxation is to broaden the line. This is a consequence of the Heisenberg uncertainty principle; if a state has a mean lifetime τ_m, then there is an uncertainty in its energy given by:

$$\Delta E = h/2\pi\tau_m \tag{3.12}$$

Since the relaxation rate is τ_m^{-1}, then the larger it is the larger ΔE will be, and hence the broader the line. Relaxation is also speeded up by interaction with nuclear quadrupole moments such as that of ^{14}N. Resonances of protons directly attached to N are often broadened.

Environmental exchange

We have already seen that the signal given by a hydroxyl proton is usually a single line, even though it is in several different magnetic environments in the medium. This happens when the rate constant for the exchange from one environment to another is greater than the frequency difference of the proton resonances in the separate environments. Thus, at one extreme, if the rate of exchange is very low, the protons will appear as separate signals, and at the other, when the rate of exchange is very fast, they appear as a single line. In between, when the rate constant of exchange is comparable to the frequency difference, we see broadened lines. Let us start with two separate signals from protons exchanging between two environments slowly, and let us imagine that the temperature is raised. Initially, we would see the separate signals broaden (Fig. 3.30a), then flatten out (Figs 3.30b and 3.30c), before they coalesce (Fig. 3.30d), and then sharpen again (Figs 3.30e and 3.30f), until they became a single narrow line. When the two environments are equally populated (Fig. 3.30a), the rate constant (s^{-1}) for the exchange at the coalescence point (Fig. 3.30d) is given by:

$$k = \pi\Delta v/\sqrt{2} \tag{3.13}$$

where Δv is the frequency separation of the initially sharp lines. Clearly NMR spectroscopy can be used to measure rate constants for those events which take place at suitable rates, often loosely referred to as the NMR time scale.

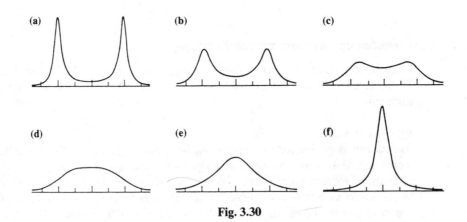

Fig. 3.30

45

For example, the ^{1}H NMR spectrum of dimethylformamide (**45**), taken at room temperature, shows two sharp singlets for the N-methyl groups at δ 2.84 and 3.0. This is because π-bonding between the nitrogen atom and the carbonyl carbon atom (**45**, arrows) slows the rotation about this bond. On warming, however, the lines broaden and coalesce, as in Fig. 3.30, with a coalescence temperature (T_c) of 337 K in a 60 MHz machine. From this observation, we can calculate the free energy of activation for the rotation using:

$$\Delta G\ddagger = RT_c\,[23+\ln(T_c/\Delta v)]$$
$$= 8.3 \times 10^{-3}\,T_c\,[23+2.3\log_{10}(T_c/\Delta v)]\ \text{kJ mol}^{-1} \qquad (3.14)$$

where T_c is expressed in Kelvin and R is the gas constant.

3.11   Improving the NMR spectrum

In several of the spectra used to illustrate this chapter, signals can be seen overlapping one another. This can be a much more serious problem when large numbers of signals overlap, and useful information is often buried in this way. For example, when there are several closely similar methylene groups in a molecule, a broad, intense, and completely unresolved signal, called the methylene envelope, is produced between δ 1 and 2. There are a few simple solutions to this problem, which we shall deal with here, and some more complicated solutions will be discussed later in the chapter.

Shift reagents
The 100 MHz spectrum of n-hexanol in CCl$_4$ is shown in Fig. 3.31a. The methyl group is resolved at high field as a distorted triplet and the methylene group adjacent to the hydroxyl group is resolved as a broad (presumably not first-order) multiplet at low field. Otherwise the carbon-bound protons are all subsumed in the methylene envelope between δ 1.2 and 1.8. The addition of a shift reagent alters this picture dramatically. Shift reagents are usually β-dicarbonyl complexes of a rare earth metal, the commonest being Eu(dpm)$_3$ (**46**), Eu(fod)$_3$ (**47**, M = Eu), and Pr(fod)$_3$ (**47**, M = Pr). These complexes are mild Lewis acids, which attach themselves to basic sites such as hydroxyl and carbonyl groups. They are also paramagnetic, and have the effect of changing substantially the magnetic field in their immediate environment. The result is a shift of the signals coming from the protons near the basic site in the organic molecule. The shift is downfield with the two europium reagents but upfield with the praseodymium reagent. Thus the spectrum of n-hexanol is spread out (Fig. 3.31b) when Eu(dpm)$_3$ is added to the solution.

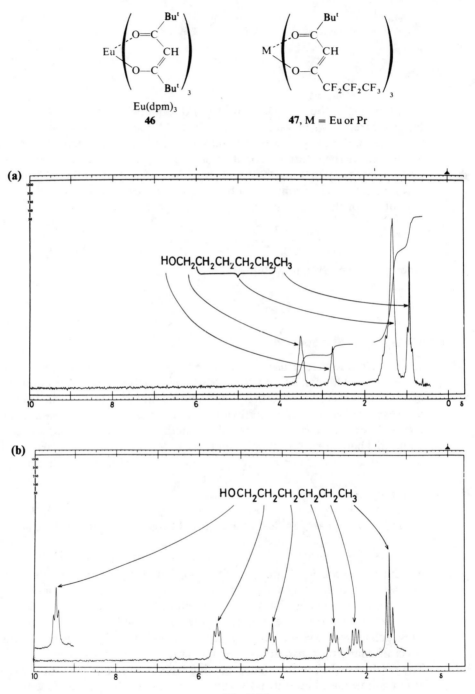

Fig. 3.31

(Superimposed trace offset 1 p.p.m.)

The shift falls off, with angular variation, as the inverse cube of the distance from the metal. The amount of shift reagent used need not be equimolar: the amount used in the spectrum illustrated was only 0.29 of an equivalent. Clearly the Lewis salt is being formed and broken rapidly on the NMR time scale, and a weighted average between the signal of the uncomplexed alcohol and the signal of the Lewis salt is detected. The penalty paid for having a paramagnetic salt present is evident in the broadness in the lines of the multiplets; fortunately the multiplicity of all the signals is still clear enough.

It is possible to add successively more and more shift reagent, and to plot the shift produced against the amount of reagent added. Table 3.32 lists the extent to which the methylene protons adjacent to some common functional groups are shifted by $Eu(dpm)_3$. This is essentially a measure of the Lewis basicity of the groups, and is a useful guide to the likely site of complexation in polyfunctional molecules, where basic groups compete for the shift reagent. Plots of this kind, carried out for each of the different kinds of proton in a single molecule, give some measure of the distance of the proton from the site of complexation, and can be useful in structure determination.

One of the most useful applications of shift reagents is in the measurement of the proportions of the R and S forms present in incompletely resolved mixtures of enantiomers. The traditional method, measurement of the rotation of polarized light, is apt to be misleading if impurities rotate the plane of the polarized light substantially more than the compound under investigation. Furthermore, this method can only be used if the extent of rotation given by one of the pure enantiomers is already known. If the chiral molecule has a basic site, and the shift reagent is a fully resolved, optically active complex such as **48**, the two enantiomers under investigation can have different binding constants and can adopt different conformations on binding, with the result that their signals are shifted to different extents. If the signals do separate adequately, they can be integrated, and the proportions of the two enantiomers measured reliably, although not very accurately.

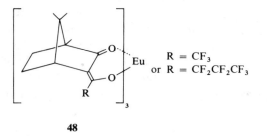

48

The effect of changing the magnetic field
NMR spectrometers are available with a variety of magnetic fields; thus 60, 80, 90, 100, 200, 250, 400, and 500 MHz machines, and several others, are in common use, with the resonance frequency for protons used to identify them. The higher field instruments have a number of substantial advantages, bringing them rapidly into general use, in spite of the cost.

The frequency of a resonance changes as the field is changed, but the chemical shift value δ does not. This means that the separation between, for example $\delta = 2$ and $\delta = 3$ is 60 Hz on a 60 MHz machine, but 250 Hz on a 250 MHz machine (remember that δ is expressed in p.p.m.). Coupling constants do not change as the magnetic field changes, but their appearance in the spectrum does. Thus a doublet with a coupling constant of 18 Hz, for example, occupies 30 per cent of the space between one δ value and another on the spectrum from a 60 MHz instrument, but only 7.2 per cent of the space on the spectrum from a 250 MHz instrument. This means that multiplets which overlap when the spectrum is taken at low field are much less likely to at high field—the multiplet is effectively narrower. Spectra taken on a high-field instrument are more likely to be first order, the signals are more easily recognized, and otherwise unresolved signals can come out of methylene envelopes. The effect can be quite dramatic, as seen in the spectra of aspirin (Fig. 3.32). Figure 3.32b shows the aromatic region measured at 60 MHz; the four different protons are all coupled to each other, and give rise to a spectrum which is not analysable, except for H_a, by the first-order approximation. Figure 3.32a shows the same part of the spectrum measured on a 250 MHz instrument, where the signals are now clearly separate and easily analysable.

A second advantage of high-field instruments is their greater sensitivity, because at higher field strengths, the numbers of nuclei, N_α and N_β, in the two spin states are more different. A third advantage is that the 'NMR time scale' changes, and the range of dynamic processes that can be studied by NMR spectroscopy is increased.

3.12   The many-pulse experiment: new techniques in FT NMR spectroscopy

The pulse and acquisition technique used in FT spectroscopy makes it possible to carry out much subtler experiments than the simple one described in Sec. 3.1. It is possible to give a second or third pulse on the same or a different axis, it is possible to wait for various lengths of time between pulses, and the pulses can be $\pi/4$ pulses, or π pulses, or any other angle, as well as the $\pi/2$ pulse illustrated in Fig. 3.2. It is not possible to describe here the pulse sequences, nor what their effect is on precessing nuclear magnets; the book by Sanders and Hunter (see the bibliography at the end of this chapter) does all that. We shall simply take the output of these experiments, illustrate them, and give a brief account of those you are most likely to encounter, so that you can recognize them, and know what kind of information can be obtained from them. The important techniques described in the rest of this chapter are rarely routine operations, and they need more instrument and computer time than regular spectra.

3.13   The separation of chemical shift and coupling onto different axes

The conventional NMR spectrum is called a one-dimensional spectrum because it has only one frequency dimension: the chemical shift and the coupling are displayed on the same axis with intensity plotted in the second dimension. With larger molecules, this can lead to very complicated spectra, with many overlapping multiplets, even at high field, and with little hope that a shift reagent will help much.

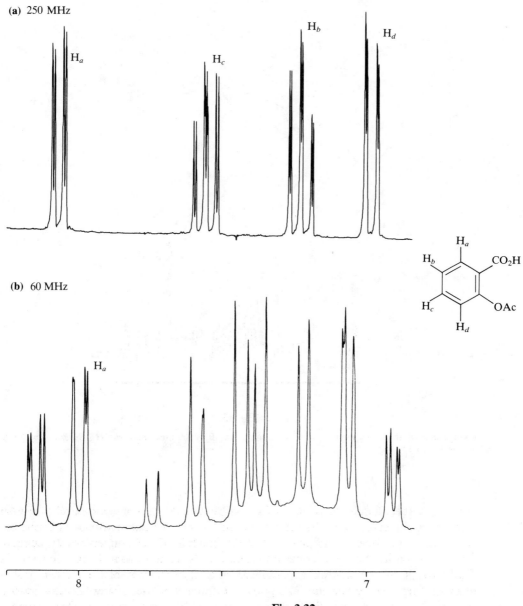

Fig. 3.32

For example, Fig. 3.33 shows the high-field region of the conventional 400 MHz spectrum of the 6-methylprogesterone (**49**), which has 24 different kinds of proton, most of them in methylene groups.

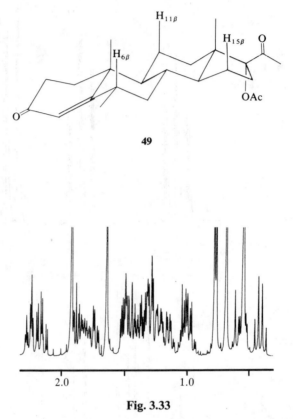

49

Fig. 3.33

(Reproduced with permission from J. K. M. Sanders and B. K. Hunter, *Modern NMR Spectroscopy*, OUP, Oxford, 1987.)

Using a special pulse sequence, it is possible to collect information and to display it in three dimensions, so that chemical shift is on the conventional axis and the coupling information is on a new axis stretching behind the conventional spectrum. This is shown for the same compound, and for the same region of the spectrum, in Fig. 3.34a in what is known as a stacked plot. A front view of the stacked plot is shown in Fig. 3.34b, which has 22 signals, one for each of the different kinds of proton (H_4 and $H_{16\beta}$ are at lower field). It is a proton NMR spectrum with all the proton–proton coupling removed, except that the signals are of different intensity from each other, because they are projections only of the tallest of the components of each multiplet.

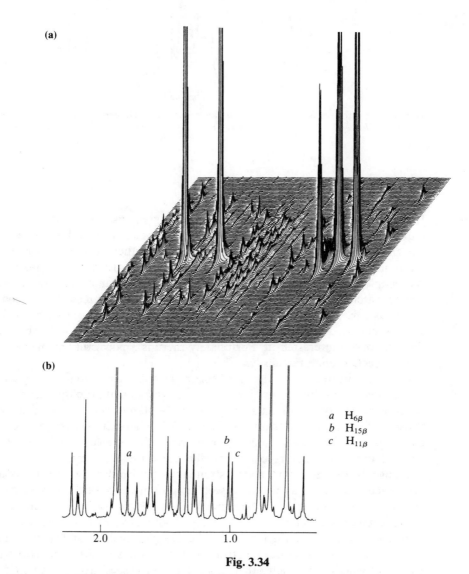

(a)

(b)

a $H_{6\beta}$
b $H_{15\beta}$
c $H_{11\beta}$

2.0   1.0

Fig. 3.34

(Reproduced with permission from J. K. M. Sanders and B. K. Hunter, *Modern NMR Spectroscopy*, OUP, Oxford, 1987.)

These signals can be plotted one at a time using the other axis, in order to display each multiplet in turn. Three examples are illustrated in Fig. 3.35, to show how signals essentially lost in Fig. 3.33 can be extricated using this technique.

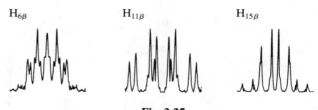

H$_{6\beta}$ H$_{11\beta}$ H$_{15\beta}$

Fig. 3.35

(Reproduced with permission from J. K. M. Sanders and B. K. Hunter, *Modern NMR Spectroscopy*, OUP, Oxford, 1987.)

3.14   Spin decoupling

Simple spin decoupling

In earlier sections we have seen how a proton gives rise to a multiplet, and how the multiplet pattern can be recognized and the coupling constants measured. However, we also want to know, if we are to gather further structural information, the multiplicity and the chemical shift of the signal from the proton or protons to which it is coupled. We have no certain way of knowing this, unless the coupling constants can be matched. For example, if the A signal of an AX system shows a coupling J_{AX} of, let us say, 5.5 Hz, the X signal must also show coupling of 5.5 Hz. This often suffices, but it requires that there be no other coupling of 5.5 Hz with which to confuse the one we are interested in. Fortunately there is a powerful technique for making this connection unambiguously.

If, during the time that a signal is being collected, the proton (or any other magnetic nucleus) has a neighbour that is exchanging its spin state rapidly, the proton we are observing will experience an average of all the states. We have seen this already in the loss of coupling to OH protons in Sec. 3.7, where the exchange was a chemical exchange, the OH protons moving from molecule to molecule with a rate constant greater than the coupling constant. The same loss of coupling occurs when the exchange is between spin states stimulated by irradiating the neighbour at its resonance frequency, as we have seen in proton-decoupled ^{13}C spectra. In those spectra the decoupling was unselective, but it is also possible to decouple selectively.

Thus, if we start with an AX system (Fig. 3.36a), determine the resonance frequency v_2 of the A signal, and then irradiate the sample precisely at that frequency at the same time as we collect the spectrum, the spin states of the A nucleus will rapidly exchange places with each other. The X nucleus 'sees' an average field from the influence of the A nucleus, and this particular coupling is effectively turned off. The X nucleus simply comes into resonance at its usual frequency v_1 as a singlet (Fig. 3.36b). When the experiment involves nuclei of different elements, it is called *heteronuclear decoupling*. When the nuclei are the same, typically both ^1H, it is called *homonuclear decoupling*. There is one limitation in homonuclear decoupling: the two signals must be reasonably well separated, or the irradiating frequency v_2 will disturb the signal at v_1. The technique works for all kinds of multiplets, so that a double doublet, for example, will collapse to a doublet when the sample is irradiated at the

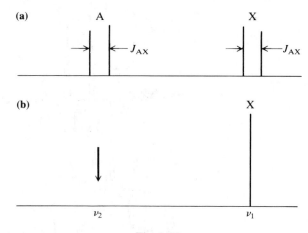

Fig. 3.36

resonance frequency of one of the protons to which it is coupled. In a complex molecule, it is possible to irradiate successively at the frequency of each of the signals in the spectrum, to plot the spectrum in each case, and to look at all the spectra to find which signals lose coupling in each experiment. In this way the full coupling relationships can be established, and much information gathered about the connections between methyl, methylene, and methine groups within a molecule. It is now more usual, however, simply to plot a COSY spectrum (see page 112) to achieve this end.

Difference decoupling

In an FT instrument, it is possible to use decoupling to reveal a buried signal. Figure 3.37c shows a narrow part of the methylene region of the spectrum of the steroid **50**. The signal is in fact a composite of the multiplets from four protons, one of them $H_{7\alpha}$, all overlapping inextricably. When the signal of $H_{6\alpha}$, which is further downfield than the ones in Fig. 3.37, is irradiated, the signal from $H_{7\alpha}$ loses one of its couplings, and the signal changes to that in Fig. 3.37b.

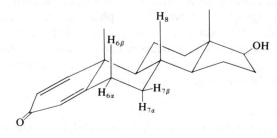

50

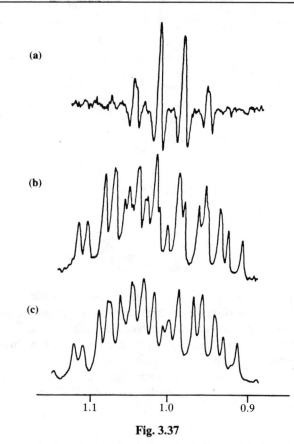

Fig. 3.37

(Reproduced with permission from J. K. M. Sanders and B. K. Hunter, *Modern NMR Spectroscopy*, OUP, Oxford, 1987.)

The multiplets are just as impossible to analyse as before, but now, because the spectrum is in the computer in digital form, it is possible to subtract the original spectrum (c) from the decoupled spectrum (b). The result is plotted in Fig. 3.37a. The other three protons were not coupled to $H_{6\alpha}$, and were unaffected by the decoupling; the subtraction therefore removed them from the signal, and left only a signal from $H_{7\alpha}$. The signal left has both the coupled and the decoupled signal in it, the original coupled signal is down and the partly decoupled signal is up, since (c) was subtracted from (b). Evidently $H_{7\alpha}$ is coupled equally to each of the protons $H_{6\beta}$, H_8, and $H_{7\beta}$, with a coupling constant of 13 Hz, leading to a quartet, and to $H_{6\alpha}$ with a coupling constant of 4.3 Hz, doubling that quartet.

COSY spectra
With suitable pulse sequences it is possible, in a single, rather extended experiment, to reveal all the coupling relationships in a molecule. This is then plotted in a three-

dimensional plot, either as a stacked plot, or, more usually, as a contour plot. The result is called a COSY spectrum (COrrelated SpectroscopY).

Figure 3.38 shows the conventional ^{1}H spectrum of *m*-dinitrobenzene and a COSY plot in the contour form for the same molecule. The lines are contours from the third dimension, representing intensity. The conventional spectrum can be seen along the diagonal, marked with a line, and the cross-peaks identify nuclei that are coupled to each other. Thus the signal from H_2 at the bottom left of the diagonal has a cross-peak connecting it (dashed lines) to the signal from $H_{4,6}$, and the signal from $H_{4,6}$ is further connected (dashed lines) by a cross-peak to the signal from H_5.

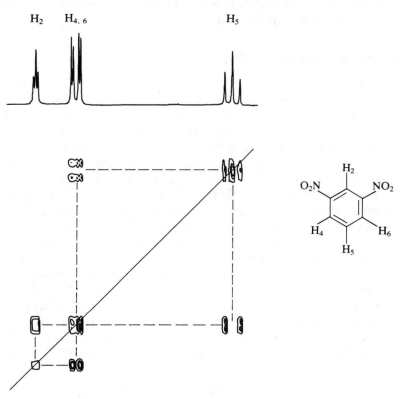

Fig. 3.38

(Reproduced with permission from J. K. M. Sanders and B. K. Hunter, *Modern NMR Spectroscopy*, OUP, Oxford, 1987.)

The cross-peaks themselves contain the coupling constants, but not the full multiplicity. Whereas the signal of H_5 on the diagonal is a triplet, viewed both from the abscissa looking up or, but less clearly, from the ordinate looking to the right, the cross-peaks do not have the centre line of the triplet in either direction. The same loss of the centre line occurs with quintets, but doublets and quartets remain as doublets and quartets in the cross-peaks. COSY spectra can be obtained to emphasize long-range coupling or to hide it. It should be obvious that the planes used to define the contours have to be chosen wisely, if all the cross-peaks are to be displayed. Sometimes as much effort has to go into preparing data for presentation as into taking spectra.

In the case of *m*-dinitrobenzene, the couplings are all evident and analysable in the conventional spectrum, but in more complicated cases it is very valuable to be able to see where all the couplings are without being obliged to carry out separate decoupling experiments for all the possible connections. Furthermore, when two (or more) signals have very similar chemical shifts, irradiation of one of the signals inevitably hits the other at the same time, and we are then unable to tell from the decoupling observed what is coupled to what. COSY spectra are not as limited in this way: as long as the signals are resolved, the cross-peaks can be associated accurately with one or the other of a closely spaced pair.

3.15   The nuclear Overhauser effect

Origins

The interaction of one magnetic nucleus with another leading to spin–spin coupling takes place through the bonds of the molecule. The information is relayed by electronic interactions, as one can see from the dependence of the coupling constant on the geometrical arrangement of the intervening bonds.

Magnetic nuclei can also interact through space, but the interaction does not lead to coupling. The interaction is revealed when one of the nuclei is irradiated at its resonance frequency and the other is detected as a more intense or weaker signal than usual. This is called *nuclear Overhauser enhancement* (NOE). The NOE effect is only noticeable over short distances, generally 2–4 Å, falling off rapidly as the inverse sixth power of the distance apart of the nuclei. The interaction is dependent upon the relaxation of the observed nucleus by the irradiated nucleus. The efficient relaxation of one nucleus by another can occur when one tumbles about the other at a frequency close to the frequency associated with the relaxation. Consider a ^{13}C nucleus and a proton close together in a molecule which rotates with a frequency of 10^9 s^{-1}. As a result, the local magnetic field at the ^{13}C nucleus will fluctuate at a rate of 10^9 s^{-1}, because the lines of force associated with the nuclear magnet of the ^{13}C nucleus will alternately reinforce (Fig. 3.39a) and subtract from (Fig. 3.39b) the applied field

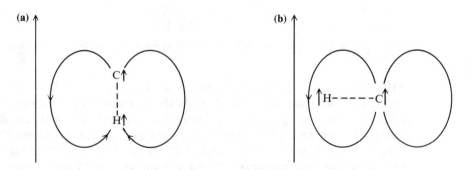

Fig. 3.39

experienced by the proton. The frequency of this field fluctuation will be close enough to the frequency associated with the $^{13}C_\alpha \rightleftharpoons {}^{13}C_\beta$ transition (usually about 10^8 s^{-1}) to promote it. The tumbling frequency is dependent upon the size of the molecule, so that relaxation is differently affected in large and in small molecules. In small molecules, with molecular weights close to 100, the tumbling frequency is close to 10^{11} s^{-1}, and the NOE is an enhancement. In large molecules, however, with a molecular weight of 1000 or more and tumbling at a frequency of approximately 10^8 s^{-1} (the precise value depending upon the viscosity of the solvent), irradiation of one signal causes the other to be weaker, rather than stronger. The effect is nevertheless still usually called an NOE, and distinction is made by referring to the former NOE as positive and the latter as negative.

Two nuclei A and X relaxing each other, but not coupling, interact to set up four populated energy levels (Fig. 3.40). Similar diagrams are set up in the full treatment of coupling, and more complicated versions of them are needed to analyse those spin interactions beyond the first order. We shall ignore the complication of coupling here, because it is quite separate from the NOE, and does not interfere. The NOE effect is through space, nuclei showing NOEs whether they are coupled or not.

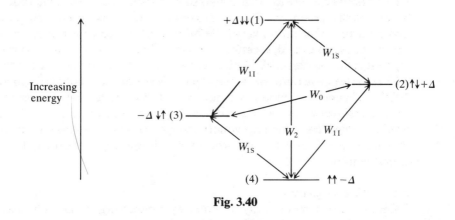

Fig. 3.40

In Fig. 3.40, the pairs of arrows depict the orientations of the nuclear magnets of the S (left) and I (right) nuclei in the applied magnetic field. The transitions W_{1S} lead to the line we associate with the S nucleus and the transitions W_{1I} lead to the line we associate with the I nucleus. If the sample is irradiated at the resonance frequency of the I nucleus, the population levels (1) and (2) grow by an amount $+\Delta$ at the expense of levels (3) and (4), respectively. There is still no obvious effect on the intensity of the S signal, because it is produced by transitions from (1) to (2) and from (3) to (4). The intensity of the S signal is dependent upon the difference between the sum of the populations of (1) and (3) and of (2) and (4), and this has not been affected. However, there are two other relaxation pathways, W_2 and W_0, which do not lead to observable signals, but do affect the populations of the four energy levels. W_2 is a two-quantum process between well-separated energy levels, and relaxation by this pathway is stimulated by the more rapid (higher frequency) tumbling of molecules with a molecular weight in the region of 100–200. The effect is to increase the population of energy level (4) at the expense of energy level (1). The sum of the populations of the energy levels (1) and (3) is reduced relative to the sum of the populations of the energy levels (2) and (4), and the signal from the S nucleus is, therefore, increased in intensity. W_0 is a zero quantum process between energy levels close in energy, and relaxation by this pathway is stimulated by the slower (lower frequency) tumbling of larger molecules with molecular weights in the region of ≥ 1000. The effect is to increase the population of energy level (3) at the expense of energy level (2). The sum of the populations in energy level (1) and (3) is now increased relative to the populations of levels (2) and (4), and the signal from the S nucleus is reduced in intensity. Molecules with intermediate molecular weight fall between two stools, and show weak or non-existent NOEs.

In ^{13}C spectra, the maximum possible NOE produced by irradiating at the proton frequency is nearly 200 per cent. NOEs of this order are found in the signals from ^{13}C atoms directly bonded to protons in proton-decoupled spectra. In ^{1}H spectra, the maximum enhancement can only be 50 per cent of the usual intensity of the signal, but the usual range is 1–20 per cent. NOEs are further weakened when the proton being observed is already being relaxed by protons other than the irradiated proton. Thus a methyl group, in which each proton already has two nearby protons to speed up relaxation, often shows very little NOE when a nearby proton is irradiated. NOEs are most easily detected, therefore, in methine groups. It is possible to measure NOEs by integrating signals with the irradiation on and off, and measuring the difference. The accuracy of integration is such that this method can only be used reliably if the NOE is 10 per cent or more, and they can only be detected in the signals from methine and, occasionally, methylene protons. Nevertheless, even in this form, NOE is a useful method for detecting which groups are close in space to each other. The technique provides valuable structural information, especially for determining stereochemistry.

NOE difference spectra

NOEs are much more easily detected by subtracting in the computer the normal spectrum from a spectrum taken with the irradiating signal on, and printing only the difference between the two spectra. All the unaffected signals simply disappear, and all that shows is the enhancement itself, together with an intense signal at the irradiating frequency. Figure 3.41 shows the complex ^{1}H spectrum of the oxindole **51**, and superimposed is the difference spectrum created after irradiating the sample at the frequency of the heavy downward-pointing arrow. This frequency is that of proton H_{7a}, which is close in space both to its neighbour H_{7b} and to $H_{5'}$ on the benzene ring. Only signals from these two protons appear in the difference spectrum, and demonstrate that the stereochemistry of the molecule is **51** and not **52**. When a similar experiment is carried out on **52**, no signal appears in the aromatic region of the difference spectrum. The signal from H_{7b} in the difference spectrum (Fig. 3.41) still shows the coupling to H_{7a}; this is because the signal used to create the NOE is applied before the acquisition pulse, but is turned off during acquisition. Coupling is therefore unaffected.

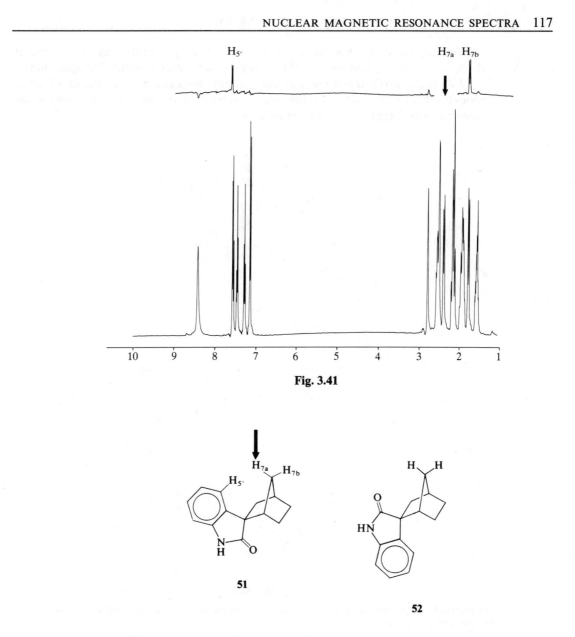

Fig. 3.41

51

52

Using difference spectra, it is easy to detect 1 per cent enhancements, or even less, with the result that NOEs in methyl groups are now quite commonly measurable. In consequence, it is usually possible to detect the NOE in both directions; not only, for example, from a methyl group to a nearby methine, but also back from the methine to the methyl group, a procedure which greatly increases one's confidence that the groups are indeed close to each other. Furthermore, the distance over which the NOE can now be detected is much greater, and more structural information is available for that reason too.

Difference NOE spectra also allow one to extract a signal from under several others, in exactly the same way as for decoupling difference spectra. The spectrum in Fig. 3.42 is the methylene region of the 6-methylprogesterone **49**, part of which is redrawn as **53**, together with the difference spectrum produced after irradiating at the resonance frequency of the C-19 methyl group.

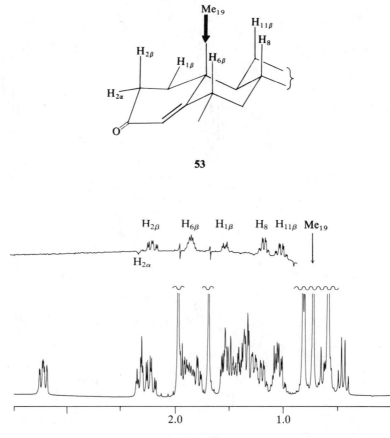

53

Fig. 3.42

(Reproduced with permission from J. K. M. Sanders and B. K. Hunter, *Modern NMR Spectroscopy*, OUP, Oxford, 1987.)

The signals in the difference spectrum come from the group of protons on the top of the molecule surrounding the 19-methyl group, each showing the multiplicity appropriate to it. In the conventional spectrum, only $H_{2\beta}$ is clearly resolved, and even that signal is more securely assigned than it was before, now that it is shown to come from a proton close in space to the C-19 methyl group. The signal from $H_{2\alpha}$ is also evident as a negative NOE. This is a common observation when a proton, $H_{2\beta}$ in this case, shows a positive NOE, and is relaying an NOE to other protons near it.

NOESY spectra

Just as the spin coupling interactions for a whole molecule can be uncovered in a COSY spectrum, so can nuclear Overhauser enhancements be collected together. The result is called a NOESY spectrum; it looks like a COSY spectrum, except that the cross-peaks are now evidence of through-space interactions instead of through-bond. The cross-peaks are most readily detected and assigned, and therefore most useful, for compounds in the molecular weight range 300–5000 Daltons. For very low molecular weight compounds, the NOE cross-peaks are inconveniently weak. As a rough guide, NOE cross-peaks indicate protons with internuclear separations lying in the range 0.2–0.4 nm, with the larger cross-peaks generally indicating a distance at the lower end of this range.

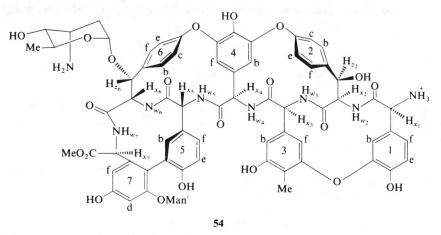

54

Figure 3.43 is a portion of the NOESY spectrum of a derivative of the antibiotic ristocetin A (**54**). The spectrum was taken in CD_3CN/D_2O as solvent, so that the NH and OH resonances are removed from the spectrum. Since the contour plot along the 45° diagonal (which represents the normal one-dimensional spectrum) is not the clearest way to follow the proton resonances, the one-dimensional spectrum is frequently reproduced (as in Fig. 3.43) along one axis of the two-dimensional contour plot. In Fig. 3.43, nine cross-peaks appear which are symmetrically placed with respect to the diagonal. The interpretation of any particular spectrum may be complicated by overlapping resonances (such as the six overlapping resonances at ~ 7.2 p.p.m. and the four at ~ 5.4 p.p.m.), but these overlaps can often be removed by running the spectrum again at different temperatures, pH, or in a modified solvent.

The data in Fig. 3.43, together with data obtained from spectra measured at other temperatures, show that the following pairs of protons are close to each other in space: $2f \leftrightarrow 2e$, $6b \leftrightarrow 6c$, $6b \leftrightarrow z_6$, $1f \leftrightarrow 1e$, $1f \leftrightarrow x_1$, $2c \leftrightarrow 4b$, $2b \leftrightarrow z_2$, $1b \leftrightarrow 3f$, and $3b \leftrightarrow x_3$. The proximities of the pairs $2f \leftrightarrow 2e$, $6b \leftrightarrow 6c$, and $1f \leftrightarrow 1e$ were, of course, already available from spin decoupling and COSY experiments. However, the NOESY experiment, by communicating information through space, allows connections to be made between the various spin systems already defined by the spin decoupling and COSY experiments. Thus the x_6,z_6 spin-coupled pair must be close in space to the aromatic ring 6, because of the existence of the NOE 6b ↔ z_6; similarly the x_2,z_2 spin coupled pair must be close to the aromatic ring 2, because of the NOE 2b ↔ z_2. Additionally, this partial spectrum allows some of the α-CH protons (x_1, x_2, ..., x_7) of the peptide backbone to be correlated with the appropriate aromatic ring (for example, 1f ↔ x_1 and 3b ↔ x_3). In a similar way, the proximity of aromatic rings in the structure is indicated (2c ↔ 4b, 1b ↔ 3f). Indeed, with only limited chemical information and a molecular weight from FAB mass spectrometry (Chapter 4), it is possible from complete COSY and NOESY spectra to solve structures as complex as **54** with complete stereochemical detail. The NOESY experiment is one of enormous power. Although it does not give quite the wealth of data and precision available from X-ray crystallography, it has the compensating advantages that it applies to molecules in solution and the compound does not have to be crystalline.

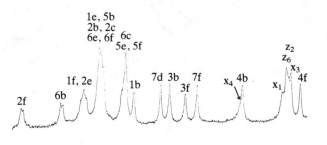

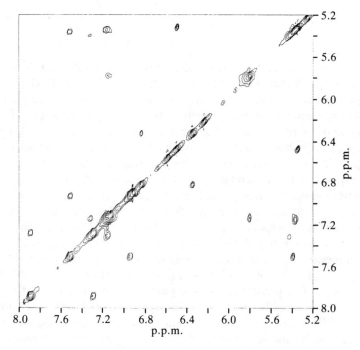

Fig. 3.43

3.16 Associating the signals from directly bonded ^{13}C and ^{1}H

Because of the lack of multiplicity in the conventional decoupled ^{13}C spectra, it is not always an easy matter to identify which carbon signals are which. Although we have seen that ^{13}C signals and ^{1}H signals are approximately equally affected in their chemical shifts by their surroundings, it is very useful in assigning ^{13}C spectra to be able to say with more certainty than simply a chemical shift argument that a particular ^{13}C signal comes from the carbon atom that carries an identifiable proton. Several multi-pulse experiments make this possible.

The first thing to know is whether the carbon atom in question is a quaternary, methine, methylene, or methyl carbon. This can be done using off-resonance decoupling (Sec. 3.6), but it can also be done in other ways. The earliest of these, called INEPT, has largely been superseded by a technique called DEPT, which displays three successive ^{13}C spectra, one containing only the methines, one only the methylenes, and one only the methyls. Figure 3.44 shows the DEPT spectra of astaxanthin **55**. The top spectrum is the full spectrum, and the subspectra below it show the signals successively of the methine, methylene, and methyl carbons.

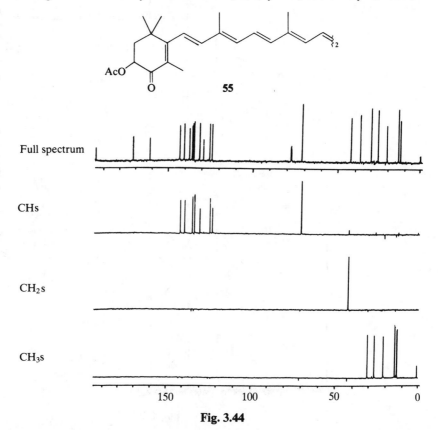

Fig. 3.44

(Reproduced with permission from J. K. M. Sanders and B. K. Hunter, *Modern NMR Spectroscopy*, OUP, Oxford, 1987.)

A second method gives even more information: it correlates the signal from the ^{13}C spectrum to the signals in the ^{1}H spectrum of the protons attached to that carbon. Figure 3.45 shows, together with the conventional spectra, the ^{1}H–^{13}C correlated spectrum of 2-butanol (**56**), where the contours of the cross-peaks identify (dashed lines) the connections. Notice that the OH signal has no connection to the carbon spectrum. These contours show no resolution here, but if each were plotted and looked at from the proton direction, in a well-resolved spectrum, they would show the proton–proton couplings. This is another way of extracting a signal from under several signals, and revealing its multiplicity.

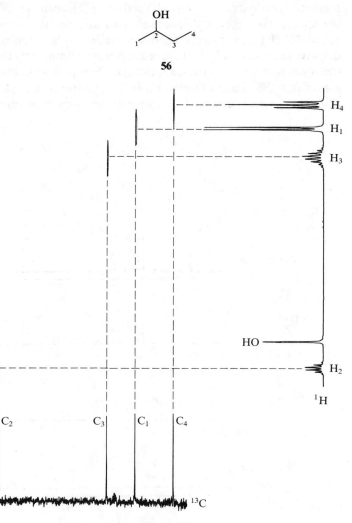

Fig. 3.45

(Reproduced with permission from J. K. M. Sanders and B. K. Hunter, *Modern NMR Spectroscopy*, OUP, Oxford, 1987.)

3.17 Identifying other connections

The spectra above show the direct connections between ^1H and ^{13}C. In assigning spectra and determining structures it can be useful to know other connections. This is especially the case when one part of a molecule is connected to another by 'spectroscopically silent' regions. For example, vicinal proton–proton coupling allows us to identify the components which make up a chain of carbon atoms. However, if a quaternary carbon intervenes, there are no protons on it to couple to the adjacent protons and the chain is interrupted. The same is true for another chain approaching the quaternary carbon from another side. We need methods for showing that the two chains are connected to the same carbon atom, and in this way build up the connections of a larger part of the molecule.

One way to do this is to find out which ^{13}C is bonded to which ^{13}C. A pulse sequence known as INADEQUATE does this, but is dependent upon detecting the signals from the minute number (1 in 10^4) of the total number of carbon atoms that are ^{13}C atoms directly bonded to other ^{13}C atoms. Nevertheless, it is a very powerful technique. Figure 3.46 shows the INADEQUATE spectrum of 2-butanol (56) in contour form, together with the conventional ^{13}C spectrum. It is best to start with an unambiguously assignable signal, which in this case we can take to be the signal of C-2, downfield because it carries a hydroxyl group. The cross-peaks labelled a and b identify the connections between C-2 and C-3 and between C-2 and C-1, respectively. The cross-peaks c then identify the remaining connection between C-3 and C-4. The dashed line bisects the mid-point between each of the pairs of cross-peaks, and is useful in picking the cross-peaks out from noise. In interpreting these spectra, a

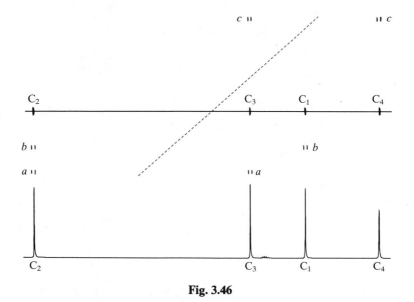

Fig. 3.46

(Reproduced with permission from J. K. M. Sanders and B. K. Hunter, *Modern NMR Spectroscopy*, OUP, Oxford, 1987.)

connectivity is initially established by making a horizontal correlation between cross-peaks symmetrically placed with respect to the dashed line, and further connectivity is established by vertical correlation at either end to other cross-peaks. Hence the termini of carbon chains are readily identified, because they only have vertical correlations at one end. Thus the upfield b and c cross-peaks have no cross-peaks above or below them, showing that these atoms are bonded only to one carbon atom each.

Another technique for making connections is to modify the method used for taking ^1H–^{13}C correlated spectra. Instead of using the one-bond couplings that identify the ^1H–^{13}C connections (Fig. 3.45), it is possible to set parameters into the pulse sequence to pick up the two- and three-bond couplings instead.

Tables of data

Table 3.4 Some parameters of magnetic nuclei

Isotope	NMR frequency (MHz) at 23.49 G	Natural abundance (%)	Relative sensitivity	Spin (I)†
^{1}H	100.00	99.98	1.00	$\frac{1}{2}$
^{2}H	15.35	1.5×10^{-2}	9.65×10^{-3}	1
^{3}H	106.7	0	1.21	$\frac{1}{2}$
^{7}Li	38.86	92.58	0.293	$\frac{3}{2}$
^{11}B	32.08	80.42	0.165	$\frac{3}{2}$
^{13}C	25.14	1.11	1.59×10^{-2}	$\frac{1}{2}$
^{14}N	7.229	99.63	1.01×10^{-3}	1
^{15}N	10.13	0.37	1.04×10^{-3}	$(-)\frac{1}{2}$
^{17}O	13.56	3.7×10^{-2}	2.91×10^{-2}	$(-)\frac{5}{2}$
^{19}F	94.08	100	0.833	$\frac{1}{2}$
^{23}Na	26.45	100	9.25×10^{-2}	$\frac{3}{2}$
^{27}Al	26.06	100	0.206	$\frac{5}{2}$
^{29}Si	19.86	4.70	7.84×10^{-3}	$(-)\frac{1}{2}$
^{31}P	40.48	100	6.63×10^{-2}	$\frac{1}{2}$
^{35}Cl	9.80	75.53	4.70×10^{-3}	$\frac{3}{2}$
^{39}K	4.67	93.10	5.08×10^{-4}	$\frac{3}{2}$
^{41}K	2.56	6.88	8.40×10^{-5}	$\frac{3}{2}$
^{51}V	26.28	99.76	0.382	$\frac{7}{2}$
^{53}Cr	5.65	9.55	9.03×10^{-4}	$\frac{3}{2}$
^{55}Mn	24.66	100	0.175	$\frac{5}{2}$
^{57}Fe	3.23	2.19	3.37×10^{-5}	$\frac{1}{2}$
^{59}Co	23.61	100	0.277	$\frac{7}{2}$
^{65}Cu	28.40	30.91	0.114	$\frac{3}{2}$
^{79}Br	25.05	50.54	7.86×10^{-2}	$\frac{3}{2}$
^{81}Br	27.00	49.46	9.85×10^{-2}	$\frac{3}{2}$
^{85}Rb	9.65	72.15	1.05×10^{-2}	$\frac{5}{2}$
^{113}Cd	22.18	12.26	1.09×10^{-2}	$\frac{1}{2}$
^{119}Sn	37.27	8.58	5.18×10^{-2}	$(-)\frac{1}{2}$
^{133}Cs	13.12	100	4.74×10^{-2}	$\frac{7}{2}$
^{195}Pt	21.50	33.8	9.94×10^{-3}	$\frac{1}{2}$
^{207}Pb	20.92	22.6	9.16×10^{-3}	$\frac{1}{2}$

† A minus sign in parentheses (−) beside an entry in the Spin I column signifies that the magnetogyric ratio, and hence magnetic moment, is negative.

Table 3.5 ^{13}C Chemical shifts in some alkanes

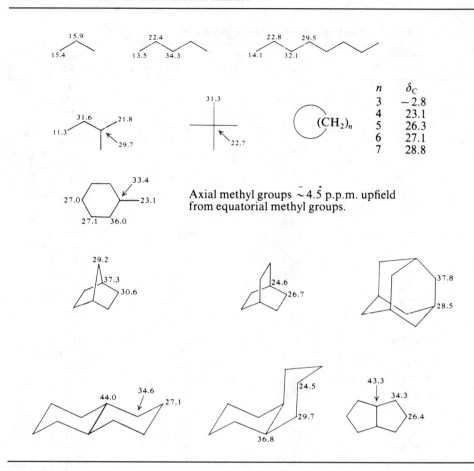

n	δ_C
3	-2.8
4	23.1
5	26.3
6	27.1
7	28.8

Axial methyl groups ~ 4.5 p.p.m. upfield from equatorial methyl groups.

Estimation of ^{13}C chemical shifts in aliphatic chains

$$\delta_C = -2.3 + \sum z + \sum S + \sum K \tag{3.15}$$

where z is the substituent constant (Table 3.6)
      S is a 'steric' correction (Table 3.7)
and K is a conformational increment for γ-substituents (Table 3.8)

Table 3.6 Substituent constants z for Eq. 3.15

	Substituent	z			
		α	β	γ	δ
C	H—	0	0	0	0
	alkyl—	9.1	9.4	−2.5	0.3
	—C=C—	19.5	6.9	−2.1	0.4
	—C≡C—	4.4	5.6	−3.4	−0.6
	Ph—	22.1	9.3	−2.6	0.3
	OHC—	29.9	−0.6	−2.7	0.0
	—CO—	22.5	3.0	−3.0	0.0
	—O_2C—	22.6	2.0	−2.8	0.0
	NC—	3.1	2.4	−3.3	−0.5
N	>N—	28.3	11.3	−5.1	0.0
	O_2N—	61.6	3.1	−4.6	−1.0
O	—O—	49.0	10.1	−6.2	0.0
	—COO—	56.5	6.5	−6.0	0.0
Hal	F—	70.1	7.8	−6.8	0.0
	Cl—	31.0	10.0	−5.1	−0.5
	Br—	18.9	11.0	−3.8	−0.7
	I—	−7.2	10.9	−1.5	−0.9
Other	—S—	10.6	11.4	−3.6	−0.4
	—SO—	31.1	9.0	−3.5	0.0

Table 3.7 'Steric' correction S for Eq. 3.15

Observed ^{13}C atom	Number of substituents other than H on the atoms directly bonded to the observed ^{13}C†			
	1	2	3	4
Primary	0.0	0.0	−1.1	−3.4
Secondary	0.0	0.0	−2.5	−7.5
Tertiary	0.0	−3.7	−9.5	−15.0
Quaternary	−1.5	−8.4	−15.0	−25.0

† Except that CO_2H, CO_2R, and NO_2 groups are counted as primary (column 1), Ph, CHO, $CONH_2$, CH_2OH, and CH_2NH_2 groups as secondary (column 2), and COR groups as tertiary (column 3).

Table 3.8 Conformational correction K for γ-substituents in Eq. 3.15

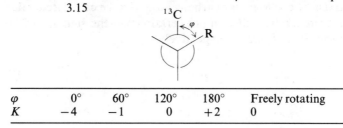

φ	0°	60°	120°	180°	Freely rotating
K	−4	−1	0	+2	0

Example of application of Eq. 3.15

The malonate ester **56** has ^{13}C signals at 13.81, 14.10, 22.4, 28.5, 29.5, 52.03, 61.12, and 169.32.

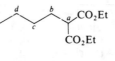

56

Take the methine carbon a:

Base value	-2.3	(methane)
1 α alkyl group	9.1	(Bu^n)
1 β alkyl group	9.4	(Pr^n)
3 γ alkyl groups	-7.5	(Et and 2 Et groups of the OEt groups)
3 δ alkyl groups	0.9	(Me and the 2 Me groups of the OEt groups)
2 α CO$_2$R groups	45.2	
S	-2.2	(a is a tertiary carbon bonded to two CO$_2$Et groups, which count as primary)
K	0	(open chain compound)
Calculated shift	52.6	Observed value 52.03

There is no difficulty in assigning this signal, because it is the only methine and is easily identified as such. However, the three methylenes at 22.4, 28.5, and 29.5 are less securely identifiable. The corresponding calculation for carbon b is:

Base value	-2.3	(methane)
2 α alkyl groups	18.2	[Pr^n and (EtO$_2$C)$_2$CH]
1 β alkyl group	9.4	(Et)
1 γ alkyl group	-2.5	(Me)
2 δ alkyl groups	0.6	(Et groups of the CO$_2$Et groups)
2 β CO$_2$R groups	4.0	
S	-2.5	(b is secondary and bonded to a carbon, namely a, with three groups on it other than hydrogen)
Calculated shift	24.9	

Similar calculations for carbons c and d, give calculated values of 29.1 and 22.8. It is therefore likely, although not certain, that the signals at 22.4, 28.5 and 29.5 can be assigned to d, b, and c, respectively.

Table 3.9 ^{13}C Chemical shifts in some alkenes and alkynes

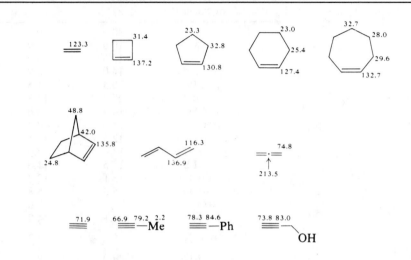

Estimation of ^{13}C chemical shifts in substituted alkenes

$$\delta_C = 123.3 + \sum z_1 + \sum z_2 + \sum S \qquad (3.16)$$

where z_1 and z_2 are the substituent constants (Table 3.10) and S is a 'steric' correction for alkyl substituents:

For each pair of *cis* substituents	$S = -1.1$
For a pair of geminal substituents on C-1	$S = -4.8$
For a pair of geminal substituents on C-2	$S = 2.5$

Table 3.10   Substituent constants z for Eq. 3.16

	Substituent R	z_1	z_2
	H—	0	0
C	Me—	10.6	−7.9
	Et—	15.5	−9.7
	Pr^n—	14.0	−8.2
	Pr^i—	20.4	−11.5
	Bu^t—	25.3	−13.3
	ClCH$_2$—	10.2	−6.0
	HOCH$_2$—	14.2	−8.4
	Me$_3$SiCH$_2$—	12.5	−12.5
	CH$_2$=CH—	13.6	−7.0
	Ph—	12.5	−11.0
	OHC—	13.1	12.7
	RCO—	15.0	5.8
	RO$_2$C—	6.3	7.0
	NC—	−15.1	14.2
N	RAcN—	6.5	−29.2
O	RO—	29.0	−39.0
	AcO—	18.4	−26.7
Hal	F—	24.9	−34.3
	Cl—	2.6	−6.1
	Br—	−7.9	−1.4
	I—	−38.1	7.0
Other	Me$_3$Si—	16.9	16.1
	RS—	18.0	−16.0
	Ph$_2$P(=O)—	8.0	11.0

Example, 2-methylbut-2-ene:

a	Base value	123.3	b	Base value	123.3
	1-methyl	10.6		2 × 1-methyl	21.2
	2 × 2-methyl	−15.8		2-methyl	−7.9
	1 *cis* pair	−1.1		1 *cis* pair	−1.1
	1 gem pair on C-2	2.5		1 gem pair on C-1	−4.8
	Calculated	119.5		Calculated	130.7
	Observed	118.5		Observed	131.8

Identifying the geometry of a disubstituted double bond

The trigonal carbons of *cis* double bonds generally come into resonance about 1 p.p.m. downfield relative to the corresponding signals from the *trans* isomers. A better test is that the allylic carbons of *cis* alkenes generally come into resonance 2–3 p.p.m. downfield relative to the same carbons in the corresponding saturated molecule, but the allylic carbons of the *trans* isomer are 2–3 p.p.m. upfield. There is, therefore, usually a difference between the *cis* and *trans* isomers of 4–6 p.p.m.

Table 3.11 ^{13}C Chemical shifts in some arenes

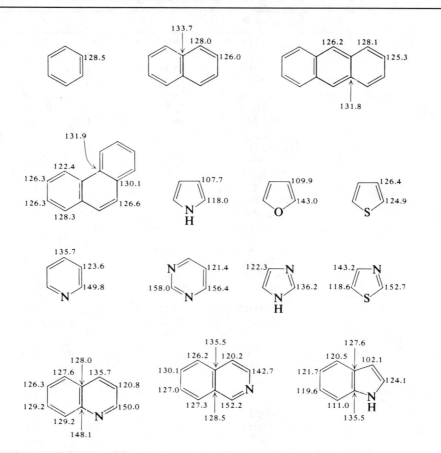

Estimation of ^{13}C chemical shifts in substituted benzenes

$$\delta_C = 128.5 + \sum z_i \qquad (3.17)$$

Table 3.12   Substituent constants z for Eq. 3.17

	Substituent R	z_1	z_2	z_3	z_4
	H—	0	0	0	0
C	Me—	9.3	0.6	0.0	−3.1
	Et—	15.7	−0.6	−0.1	−2.8
	Pr^n—	14.2	−0.2	−0.2	−2.8
	Pr^i—	20.1	−2.0	0.0	−2.5
	Bu^t—	22.1	−3.4	−0.4	−3.1
	$ClCH_2$—	9.1	0.0	0.2	−0.2
	$HOCH_2$—	13.0	−1.4	0.0	−1.2
	$CH_2{=}CH$—	7.6	−1.8	−1.8	−3.5
	Ph—	13.0	−1.1	0.5	−1.0
	$HC{\equiv}C$—	−6.1	3.8	0.4	−0.2
	OHC—	9.0	1.2	1.2	6.0
	MeCO—	9.3	0.2	0.2	4.2
	RO_2C—	2.1	1.2	0.0	4.4
	NC—	−16.0	3.5	0.7	4.3
N	H_2N—	19.2	−12.4	1.3	−9.5
	Me_2N—	22.4	−15.7	0.8	−11.8
	AcNH—	11.1	−16.5	0.5	−9.6
	O_2N—	19.6	−5.3	0.8	6.0
O	HO—	26.9	−12.7	1.4	−7.3
	MeO—	30.2	−14.7	0.9	−8.1
	AcO—	23.0	−6.4	1.3	−2.3
Hal	F—	35.1	−14.3	0.9	−4.4
	Cl—	6.4	0.2	1.0	−2.0
	Br—	−5.4	3.3	2.2	−1.0
	I—	−32.3	9.9	2.6	−0.4
Other	Me_3Si—	13.4	4.4	−1.1	−1.1
	Ph_2P—	8.7	5.1	−0.1	0.0
	MeS—	9.9	−2.0	0.1	−3.7

Table 3.13 ^{13}C Chemical shifts of carbonyl carbons

R^1	R^2	δ_C		R^1	R^2	δ_C
Me—	—H	199.7		Me—	—OH	178.1
Et—	—H	206.0		Et—	—OH	180.4
Pr^i—	—H	204.0		Pr^i—	—OH	184.1
CH_2=CH—	—H	192.4		Bu^t—	—OH	185.9
Ph—	—H	192.0		CH_2=CH—	—OH	171.7
				Ph—	—OH	172.6
Me—	—Me	206.0				
Et—	—Me	207.6		Me—	—OMe	170.7
Pr^i—	—Me	211.8		Et—	—OMe	173.3
Bu^t—	—Me	213.5		Pr^i—	—OMe	175.7
$ClCH_2$—	—Me	200.7		Bu^t—	—OMe	178.9
Cl_2CH—	—Me	193.6		CH_2=CH—	—OMe	165.5
Cl_3C—	—Me	186.3		Ph—	—OMe	166.8
CH_2=CH—	—Me	197.2		$-(CH_2)_3O-$		177.9
Ph—	—Me	197.6		$-(CH_2)_4O-$		175.2
$-(CH_2)_3-$		208.2				
$-(CH_2)_4-$		213.9		Me—	—NH_2	172.7
$-(CH_2)_5-$		208.8		CH_2=CH—	—NH_2	168.3
$-(CH_2)_6-$		211.7		Ph—	—NH_2	169.7
cyclopentenone		209.0		$-(CH_2)_3NH-$		179.4
				$-(CH_2)_4NH-$		173.0
				Me—	—OAc	167.3
				Ph—	—OAc	162.8
cyclohexenone		198.0		Me—	—Cl	168.6
				CH_2=CH—	—Cl	165.6
				Ph—	—Cl	168.0

Table 3.14 $^{13}C-^{1}H$ Coupling constants

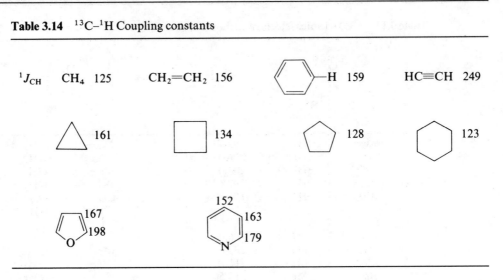

Estimation of $^{1}J_{CH}$ in alkanes

$$\text{For } R^{1}R^{2}R^{3}CH, \qquad ^{1}J_{CH} = 125 + \sum z_{i} \qquad (3.18)$$

Table 3.15   Substituent constants for Eq. 3.18

	Substituent R^{i}	z		Substituent R^{i}	z
	H—	0	N	H_2N—	8
C	Me—	1	O	HO—	18
	Bu^{t}—	−3	Hal	F—	24
	$ClCH_2$—	3		Cl—	27
	HC≡C—	7		Br—	27
	Ph—	1		I—	26
	OHC—	2			
	MeO_2C—	−1			
	NC—	11			

Table 3.16 $^{13}C-F$ Coupling constants

$RCH_2CH_2CH_2F$ $^{1}J_{CF}$ 165 $^{2}J_{CF}$ 18 $^{3}J_{CF}$ 6 Hz

$^{1}J_{CF}$ 245 $^{2}J_{CF}$ 21 $^{3}J_{CF}$ 8 $^{4}J_{CF}$ 3 Hz

Table 3.17 ^{1}H Chemical shifts in methyl, methylene, and methine groups

	Methyl protons	δ_H	Methylene protons	δ_H	Methine protons	δ_H
C	CH_3—R	0.9	R—CH_2—R	1.4	$>$CH—R	1.5
	CH_3—C—C=C	1.1	R—CH_2—C—C=C	1.7		
	CH_3—C—O	1.3	R—CH_2—C—O	1.9	$>$CH—C—O	2.0
	CH_3—C—N	1.1	R—CH_2—C—N	1.4		
	CH_3—C—NO_2	1.6	R—CH_2—C—NO_2	2.1		
	CH_3—C=C	1.6	R—CH_2—C=C	2.3		
	CH_3—Ar	2.3	R—CH_2—Ar	2.7	$>$CH—Ar	3.0
	CH_3—C=CC=O	2.0	R—CH_2—C=CC=O	2.4		
	C=C(CH_3)—C=O	1.8	C=C(CH_2—R)—C=O	2.4		
	CH_3—C≡C	1.8	R—CH_2—C≡C	2.2	$>$CH—C≡C	2.6
	CH_3—CO—R	2.2	R—CH_2—CO—R	2.4	$>$CH—CO—R	2.7
	CH_3—CO—Ar	2.6	R—CH_2—CO—Ar	2.9	$>$CH—CO—Ar	3.3
	CH_3—CO—OR	2.0	R—CH_2—CO—OR	2.2	$>$CH—CO—OR	2.5
	CH_3—CO—OAr	2.4				
	CH_3—CO—N	2.0	R—CH_2—CO—N	2.2	$>$CH—CO—N	2.4
			R—CH_2—C≡N	2.3	$>$CH—C≡N	2.7
N	CH_3—N	2.3	R—CH_2—N	2.5	$>$CH—N	2.8
	CH_3—N—Ar	3.0				
	CH_3—N—CO—R	2.9	R—CH_2—N—CO—R	3.2	$>$CH—N—CO—R	4.0
	CH_3—N$^+$	3.3	R—CH_2—N$^+$	3.3		
			R—CH_2—NO_2	4.4	$>$CH—NO_2	4.7
O			R—CH_2—OH	3.6	$>$CH—OH	3.9
	CH_3—OR	3.3	R—CH_2—OR	3.4	$>$CH—OR	3.7
	CH_3—O—C=C	3.8	R—CH_2—O—C=C	3.7		
	CH_3—OAr	3.8	R—CH_2—OAr	4.3	$>$CH—OAr	4.5
	CH_3—O—CO—R	3.7	R—CH_2—O—CO—R	4.1	$>$CH—O—CO—R	4.8
			RO—CH_2—OR	4.8		
Hal			R—CH_2—F	4.4		
			R—CH_2—Cl	3.6	$>$CH—Cl	4.2
			R—CH_2—Br	3.5	$>$CH—Br	4.3
			R—CH_2—I	3.2	$>$CH—I	4.3
Other	CH_3—Si	0.0	R—CH_2—Si	0.5	$>$CH—Si	1.2
	CH_3—S	2.1	R—CH_2—S	2.4	$>$CH—S	3.2
	CH_3—S(O)R	2.5				
	CH_3—S(O_2)R	2.8	R—CH_2—S(O_2)R	2.9		
			RS—CH_2—SR	4.2		

R = alkyl group. These values will usually be within ± 0.2 p.p.m. unless electronic or anisotropic effects from other groups are strong. An obsolete scale used τ values; these are related to δ values by the simple equation $\tau = 10 - \delta$.

Estimation of ^1H chemical shifts in substituted alkanes

$$R^1R^2R^3CH \qquad \delta_H = 1.50 + \sum z_i \qquad\qquad (3.19)$$

Table 3.18   Substituent constants z for Eq. 3.19

R^i	z	R^i	z	R^i	z
H—	−0.3	HC≡C—	0.9	MeO—	1.5
alkyl—	0.0	OHC—	1.2	PhO—	2.3
CH$_2$=CHCH$_2$—	0.2	MeCO—	1.2	AcO	2.7
MeCOCH$_2$—	0.2	RO$_2$C—	0.8	Cl—	2.0
HOCH$_2$—	0.3	NC—	1.2	Br—	1.9
ClCH$_2$—	0.5	H$_2$N—	1.0	I—	1.4
CH$_2$=CH—	0.8	O$_2$N—	3.0	MeS—	1.0
Ph—	1.3	HO—	1.7	Me$_3$Si—	−0.7

Table 3.19 ^1H Chemical shifts of methylene groups in some cyclic compounds

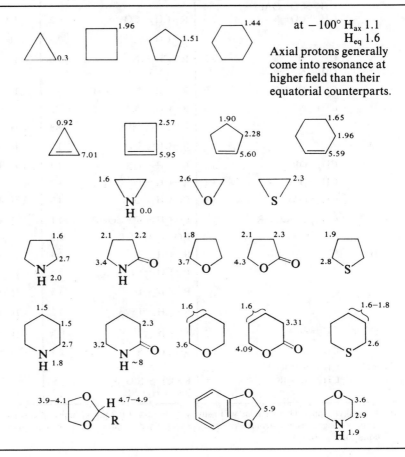

at −100° H$_{ax}$ 1.1
H$_{eq}$ 1.6
Axial protons generally
come into resonance at
higher field than their
equatorial counterparts.

Table 3.20 ^{1}H Chemical shifts of protons attached to multiple bonds

Structure	δ_H	Structure	δ_H
RCHO	9.4–10.0	$>$C=CH—	4.5–6.0
ArCHO	9.7–10.5	$>$C=CHCO—	5.8–6.7
—OCHO	8.0–8.2	—HC=CCO—	6.5–8.0
$>$NCHO	8.0–8.2	—HC=C—O—	4.0–5.0
—C≡CH	1.8–3.1	$>$C=CH—O—	6.0–8.1
$>$C=C=CH—	4.0–5.0	—HC=C—N—	3.7–5.0
ArH	6.0–9.0	$>$C=CH—N—	5.7–8.0

Estimation of ^{1}H chemical shift in alkenes

$$\delta_H = 5.25 + z_{gem} + z_{cis} + z_{trans} \qquad (3.20)$$

Table 3.21 Substituent constants z for Eq. 3.20

	R	z_{gem}	z_{cis}	z_{trans}
	H—	0	0	0
C	alkyl—	0.45	−0.22	−0.28
	ring-alkyl—	0.69	−0.25	−0.28
	CO—CH$_2$— or NC—CH$_2$—	0.69	−0.08	−0.06
	Ar—CH$_2$—	1.05	−0.29	−0.32
	N—CH$_2$—	0.58	−0.10	−0.08
	O—CH$_2$—	0.64	−0.10	−0.02
	Hal-CH$_2$—	0.70	0.11	−0.04
	S—CH$_2$—	0.71	−0.13	−0.22
	isolated C=C—	1.00	−0.09	−0.23
	conjugated C=C—	1.24	0.02	−0.05
	Ar—	1.38	0.36	−0.07
	OHC—	1.02	0.95	1.17
	isolated RCO—	1.10	1.12	0.87
	conjugated RCO—	1.06	0.91	0.74
	isolated HO$_2$C—	0.97	1.41	0.71
	conjugated HO$_2$C—	0.80	0.98	0.32
	isolated RO$_2$C—	0.80	1.18	0.55
	conjugated RO$_2$C—	0.78	1.01	0.46
	N—CO—	1.37	0.98	0.46
	Cl—CO—	1.11	1.46	1.01
	—C≡C—	0.47	0.38	0.12
	N≡C—	0.27	0.75	0.55
N	alkyl-N—	0.80	−1.26	−1.21
	conjugated alkyl or aryl-N—	1.17	−0.53	−0.99
	—CO—N—	2.08	−0.57	−0.72
	O$_2$N—	1.87	1.30	0.62
O	alkyl-O—	1.22	−1.07	−1.21
	conjugated alkyl or aryl-O—	1.21	−0.60	−1.00
	—CO—O—	2.11	−0.35	−0.64
Hal	F—	1.54	−0.40	−1.02
	Cl—	1.08	0.18	0.13
	Br—	1.07	0.45	0.55
	I—	1.14	0.81	0.88
Other	R$_3$Si—	0.90	0.90	0.60
	RS—	1.11	−0.29	−0.13
	RSO—	1.27	0.67	0.41
	RSO$_2$—	1.55	1.16	0.93

Use the 'conjugated' values when either the substituent or the double bond is further conjugated. Use the 'ring-alkyl' values when the double bond and the alkyl group are part of a five- or six-membered ring.

Table 3.22 ^{1}H Chemical shifts of protons attached to double bonds in some unsaturated cyclic systems†

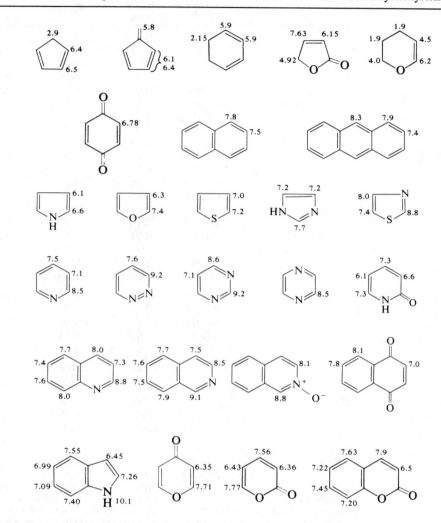

† For simple cycloalkenes, see Table 3.19.

Estimation of proton chemical shifts in substituted benzenes

$$\delta_H = 7.27 + \sum z_i \qquad (3.21)$$

Table 3.23   Substituent constants for Eq. 3.21

	R	z_{ortho}	z_{meta}	z_{para}
	H—	0	0	0
C	Me—	−0.20	−0.12	−0.22
	Et—	−0.14	−0.06	−0.17
	Pr^i—	−0.13	−0.08	−0.18
	Bu^t—	0.02	−0.08	−0.21
	H_2NCH_2— or $HOCH_2$—	−0.07	−0.07	−0.07
	$ClCH_2$—	0.00	0.00	0.00
	F_3C—	0.32	0.14	0.20
	Cl_3C—	0.64	0.13	0.10
	$CH_2{=}CH$—	0.06	−0.03	−0.10
	Ph—	0.37	0.20	0.10
	OHC—	0.56	0.22	0.29
	MeCO—	0.62	0.14	0.21
	H_2NCO—	0.61	0.10	0.17
	HO_2C—	0.85	0.18	0.27
	MeO_2C—	0.71	0.1	0.21
	ClCO—	0.84	0.22	0.36
	HC≡C—	0.15	−0.02	−0.01
	N≡C—	0.36	0.18	0.28
N	H_2N—	−0.75	−0.25	−0.65
	Me_2N—	−0.66	−0.18	−0.67
	AcNH—	0.12	−0.07	−0.28
	O_2N—	0.95	0.26	0.38
O	HO—	−0.56	−0.12	−0.45
	MeO—	−0.48	−0.09	−0.44
	AcO—	−0.25	0.03	−0.13
Hal	F—	−0.26	0.00	−0.04
	Cl—	0.03	−0.02	−0.09
	Br—	0.18	−0.08	−0.04
	I—	0.39	−0.21	0.00
Other	Me_3Si—	0.22	−0.02	−0.02
	$(MeO)_2P({=}O)$—	0.48	0.16	0.24
	MeS—	0.37	0.20	0.10

These parameters are simply the shifts measured on the corresponding monosubstituted benzene ring; they are not accurately taken over to polysubstituted benzenes, but the estimation of chemical shift is usually fairly good. Errors are particularly likely to occur when substituents *ortho* to one another interfere with conjugation to the ring.

Table 3.24 ^{1}H Chemical shifts of protons attached to elements other than carbon

	Structure	δ_H		Structure	δ_H
NH	RNH$_2$ and R$_2$NH	0.5–4.5	OH	monomeric H$_2$O	~1.5
	ArNH$_2$ and ArNHR	3–6		suspended HOD	~4.7
	RCONH$_2$ and RCONHR	5–12		ROH	0.5–4.5
	pyrrole NH	7–12		ArOH	4.5–10
				RCO$_2$H	9–13
SiH	$>$SiH	~3.8	$>$C=N—OH		9–12
SH	RSH	1–2			7–16
	ArSH	3–4			

These values are very sensitive to temperature, solvent, and concentration: the stronger the hydrogen bonding, the lower field the chemical shift.

Table 3.25 ^{13}C and residual ^{1}H chemical shifts in the common deuterated solvents

Solvent	Deuterated solvent				Undeuterated solvent
	δ_H†	Multi-plicity‡	δ_C	Multi-plicity‡	δ_C
Acetic acid	2.05				21.1
	11.5§				178.1
Acetone	2.05	quintet	29.8	septet	30.5
			205.7		205.4
Acetonitrile	1.95	quintet	1.2	septet	1.6
			117.8		117.8
Benzene	7.3		128.0	triplet	128.5
t-Butanol	1.28¶				
Carbon disulphide					192.8
Carbon tetrachloride					96.1
Chloroform	7.25		77.0	triplet	77.2
Cyclohexane	1.40	triplet	26.3	quintet	27.6
Water	4.7§				
Dimethylformamide (DMF)	2.75	quintet			
	2.95	quintet			
	8.05	triplet			
Dimethylsulphoxide (DMSO)	2.5	quintet	39.7	septet	40.6
water in DMSO	3.3§				
Dioxan	3.55	triplet			67.3
Hexamethylphosphoramide (HMPA)	2.60	double†† quintet			
Methanol	3.35	quintet	49.0	septet	49.9
	4.8§				
Dichloromethane (methylene dichloride)	5.35	triplet			54.0
Pyridine	7.0		123.4	triplet	123.9
	7.35		135.3	triplet	135.9
	8.5		149.8	triplet	150.3
Toluene	2.3	quintet			
	7.2				
Trifluoroacetic acid (TFA)	11.3§				115.7‡‡
					163.8§§

† Residual protons in the deuterated solvent.
‡ A singlet unless otherwise stated.
§ Variable, depents upon the solvent and its concentration.
¶ (CH$_3$)$_3$COD is usually used, not the fully deuterated solvent.
†† Coupling to P, $J = 9$ Hz.
‡‡ Quartet from coupling to F, $J = 294$ Hz.
§§ Quartet from coupling to F, $J = 46$ Hz.

Table 3.26   Geminal ($^2J_{HH}$) coupling constants (Hz)

R^1	R^2	$^2J_{HH}$		R^1	R^2	$^2J_{HH}$
H	H	-12.4		H	CN	-16.2
R	R	$-8 \ldots -18$		H	COMe	-14.9
$-(CH_2)_2-$		$-3 \ldots -9$				
$-(CH_2)_3-$		$-11 \ldots -17$				-21.5
$-(CH_2)_4-$		$-8 \ldots -18$				
$-(CH_2)_5-$		$-11 \ldots -14$				
H	Ph	-14.3				$-3 \ldots +3$
H	OH	-10.8				$-8 \ldots -10$
H	Cl	-10.8				
$-O(CH_2)_2O-$		~ 0				
$-O(CH_2)_3O-$		$-5 \ldots -6$				

Table 3.27   Vicinal ($^3J_{HH}$) coupling constants in some aliphatic compounds (Hz)

Open chain compounds			Cyclic compounds		
Structure	$^3J_{HH}$ range	Typical value	Structure	Ring size	$^3J_{HH}$ range
CH_3-CH_2-	6–8	7	cis	3	7–13
			trans	3	4–9.5
$CH_3-CH{<}$	5–7	6	cis	4	4–12
			trans	4	2–10
			cis	5	5–10
$-CH_2-CH_2-$	5–8	7	trans	5	5–10
			cis	6	8–13
${>}CH-CH{<}$	0–8	7	trans	6	2–6†
${>}C{=}CH-CH{<}$	4–11	6		3	1.8‡
				4	−0.8‡
${>}C{=}CH-CH{=}C{<}$	6–13	11§		5	0.5‡
				6	1.5‡
${>}CH-CHO$	0–3	2		7	3.7‡
				8	5.3‡
${>}C{=}CH-CHO$	5–8	7		3	0.5–2
				4	2.5–4
cis-$CH{=}CH-$	0–12	8		5	5–7
				6	8.5–10.5
				7	9–12.5
trans-$CH{=}CH-$	12–18	15		8	10–13
			H^7, H^1, H^{2x}, H^{3x}, H^{2n}, H^{3n}	1–2x	3–4
				1–2n	0–2
				2x–3x	9–10
				2n–3n	6–7
				2x–3n	2–5
				1–7	0–3

† $J_{aa} = 8$–13, $J_{ee} = 2$–5; note that J_{ee} is usually 1 Hz smaller than J_{ae}.
‡ Value for the unsubstituted cycloalkene.
§ Found in dienes adopting the s-trans conformation.

Table 3.28   Vicinal ($^3J_{HH}$) coupling constants (Hz) in some heterocyclic and aromatic compounds

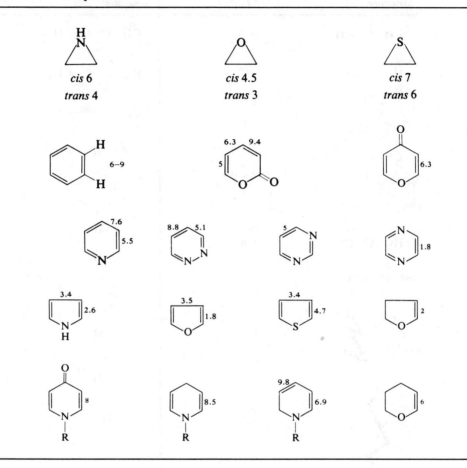

Table 3.29 Long-range ($^4J_{HH}$ and $^5J_{HH}$) coupling constants (Hz)

Structure	$^4J_{HH}$	Structure	$^5J_{HH}$
$-CH=C-CH\diagup$	0–3	$\diagdown CH-C=C-CH\diagdown$	0–2
	1–3	$-HC=C=C-CH\diagdown$	2–3
	0.6–0.9	$\diagdown CH-C\equiv C-CH\diagdown$	1–3
$-HC=C=CH-$	4–6		8–10
$HC\equiv C-CH\diagup$	1–3		
	1–2		0–1
	7–8		1–1.5
	7a–2n 3–4 2x–6x 1–2		
	signal perceptibly broadened by ^4J coupling		

Table 3.30 ^{1}H–^{19}F coupling constants (Hz)

Structure		J	Structure		J
$^2J_{HF}$	H F (isopropyl)	45–52			
	H F (cyclopropyl)	60–65	F-phenyl–H	ortho 6–11 meta 3–9 para 0–4	
	H F (vinyl)	72–90			
$^3J_{HF}$	CH_3—CF<	20–24			
	>CH—CF<	0–45†			
	cis-HC=CF—	3–20	CH_3-phenyl–F	ortho 2.5 meta 1.5 para 0	
	trans-HC=CF—	12–53			
$^4J_{HF}$	>HC—C—CF<	0–9‡			
	cis-FC=C—CH<	2–4			
	trans-FC=C—CH<	0–6			

† 0–12 when gauche and 10–45 when anti-periplanar.
‡ The higher end of the range (≥ 3.5) when the atoms are held in a W conformation.

Table 3.31 $^{31}P-^1H$ coupling constants (Hz)†

Type of coupling	Class of compound		
	Phosphines	Phosphonium salts	Phosphine oxides
$^1J_{PH}$	(150) 185–220 (250)	400–900	200–750
$^2J_{PH}$	(−5) 0–15 (27)	(0) 10–18	5–25
	46‡	30‡	40‡
$^3J_{PCCH}$	(10) 13–17 (20)	(0) 10–20 (57)	14–30
$^3J_{PC=CH}$	trans (5) 12–41	trans 28–50 (80)	
	cis§ 6–20	cis§ 10–20 (35)	
		Phosphites	Phosphates
$^3J_{POCH}$		(0) 5–14 (20)	(0) 5–20 (30)
	All compounds		
$^4J_{PH}$	0–3 (5)¶		

† The coupling constants are often strongly dependent upon the groups attached to phosphorus, and therefore values outside the quoted ranges may occasionally be observed; values in parentheses are 'extreme' values so far reported.

$$\overset{\text{C}}{\underset{\|}{}}$$

‡ Values observed in P—C—H systems.
§ *Trans* coupling is usually about twice that of *cis* coupling.
¶ In the system P—C≡C—CH.

Table 3.32   Eu(dpm)$_3$-induced shifts of protons in some common environments†

Functional group	Shift [p.p.m./mol of Eu(dpm)$_3$ per mol of substrate]
RCH$_2$NH_2	~150
RCH$_2$OH	~100
RCH_2NH$_2$	30–40
RCH_2OH	20–25
RCH_2COR′	10–17
RCH$_2$CHO	19
RCH_2CHO	11
RCH_2OCH_2R	10
RCH$_2$CO$_2$CH$_3$	7
RCH$_2$CO$_2$CH_3	6.5
RCH_2CN	3–7

† The shifts refer to the protons indicated in italics.

Bibliography

TEXTBOOKS

R. J. Abraham and P. Loftus, *Proton and Carbon-13 NMR Spectroscopy*, Heyden, London, 1978.

R. R. Ernst, G. Bodenhausen, and A. Wokaun, *Principles of Nuclear Magnetic Resonance in One and Two Dimensions*, Clarendon Press, Oxford, 1987.

T. C. Farrar and E. D. Becker, *Pulse and Fourier Transform NMR*, Academic Press, New York, 1971.

L. M. Jackman and S. Sternhell, *Applications of NMR Spectroscopy in Organic Chemistry*, Pergamon Press, London, 1969.

G. C. Levy and G. L. Nelson, *Carbon-13 NMR for Organic Chemists*, Wiley-Interscience, New York, 1972.

J. K. M. Sanders and B. K. Hunter, *Modern NMR Spectroscopy*, OUP, Oxford, 1987.

D. Shaw, *Fourier Transform NMR Spectroscopy*, Elsevier, Amsterdam, 2nd Ed., 1984.

J. B. Stothers, *Carbon-13 NMR Spectroscopy*, Academic Press, New York, 1973.

F. W. Wehrli and T. Wirthlin, *Interpretation of Carbon-13 NMR Spectra*, Heyden, London, 1976.

K. Wüthrich, *NMR of Proteins and Nucleic Acids*, Wiley, New York, 1986.

CATALOGUES

N. S. Bhacca, L. F. Johnson, and J. N. Shoolery, *High Resolution NMR Spectra Catalogue*, Vols I and II, Varian Associates, Palo Alto, 1963.

C. J. Pouchert, *The Aldrich Library of NMR Spectra*, Vols 1 and 2, Aldrich Chemical Co. Inc., 2nd Ed., 1983.

E. Pretsch, T. Clerc, J. Seibl, and W. Simon, *Tables of Spectral Data for Structure Determination of Organic Compounds*, Springer, Berlin, English Ed., 1983.

Sadtler Handbook of Proton NMR Spectra, Heyden, London.

Sadtler Guide to Carbon 13 NMR Spectra, Heyden, London.

4. Mass spectra

4.1 *Introduction*. 4.2 *Ion production*. 4.3 *Ion analysis*. 4.4 *Ion abundances in mass spectra*. 4.5 *Gas chromatography–mass spectrometry (GC/MS)*. 4.6 *Liquid chromatography–mass spectrometry LC/MS*. 4.7 *MS data systems*. 4.8 *Specific ion monitoring and quantitative MS*. 4.9 *Interpreting the spectrum of an unknown*. *Bibliography*.

4.1 Introduction

In its simplest form, the mass spectrometer is designed to perform three basic functions: to vaporize compounds of widely varying volatility; to produce ions from the resulting gas-phase molecules (except where the volatilization process directly produces ions rather than neutrals); and to separate ions according to their mass-to-charge ratios (m/ze), and subsequently detect and record them. Since multiply charged ions are produced only rarely relative to singly charged ions, z can normally be taken as one; and since e is a constant (the charge of one electron), m/z then gives the mass of the ion. Thus, the mass spectrometer is a device for the production and weighing of ions.

The devices which are used to produce gas-phase ions almost always put sufficient vibrational energy into the ion that it will, to some extent, fragment to produce new ions with loss of neutrals, e.g.

$$A^+ \nearrow \begin{array}{l} B^+ + \text{neutral} \\ \searrow C^+ + \text{neutral} \end{array}$$

Given sufficient vibrational energy, B^+ and/or C^+ may decompose further, e.g.

$$C^+ \longrightarrow D^+ + \text{neutral}$$

When the array of ions (A^+, B^+, C^+, etc.) has been separated and recorded, the output is known as a mass spectrum. It is a record of the abundance of each ion (plotted vertically) against its m/z value (plotted horizontally). The most abundant ion is arbitrarily assigned a value of 100 per cent (Fig. 4.1). Obviously, if A^+ is produced by simple ionization of the sample, then the mass of A^+ gives the molecular weight of the sample. Negative-ion spectra, although less commonly used than positive-ion spectra, can also be obtained.

It can be seen that the mass spectrum is a result of a series of competing and consecutive unimolecular reactions. Therefore, it is determined by chemical reactivity and is not a true spectroscopic method. However, since it complements information provided to the organic chemist by UV, IR, and NMR, it is conveniently considered alongside them.

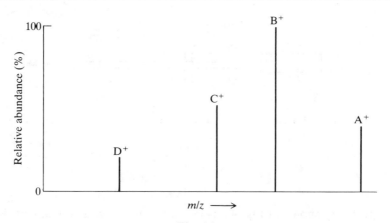

Fig. 4.1

This chapter deals first with the production of gas-phase ions, and second with their separation. In the third part, the factors which determine ion abundances in various types of spectra are given, along with some applications. Last, we consider the coupling of chromatographic devices—gas chromatographs (GC) and liquid chromatographs (LC)—to mass spectrometers, and the use of on-line computers to handle the large amounts of data which are generated.

4.2   Ion production

Volatile materials

The ions which are produced inside the mass spectrometer must, during the process of their separation according to their m/z ratio (ion analysis), be able to travel distances of the order of metres without undergoing collisions. Thus, ion analysers typically operate at pressures $\leqslant 10^{-4}$ N m^{-2} ($\leqslant 10^{-6}$ mmHg). Since the ions normally pass quickly from where they are produced to the ion analyser, this places limits upon the operating pressures (10^{-4}–10^2 N m^{-2}) where ions are produced. Thermally stable organic molecules can often be heated to 200–300°C at these pressures without significant decomposition. If such materials lack, or have very few, polar functional groups (e.g. —OH, —COOH, —NH$_2$) they will often pass into the gas phase if their molecular weights lie in the range up to about 1000 Daltons. As a useful guide, thermal volatilization of the sample is often successful for relatively non-polar molecules up to molecular weights of 1000 Daltons, or medium polarity molecules up to molecular weights of 300 Daltons. A common device to achieve such volatilization of solids is the probe shown in Fig. 4.2, where the probe tip can be heated. Alternatively, readily volatile samples (e.g. some relatively low molecular weight solids and most liquids) can be volatilized by heating under low pressure at a site remote from the ion production region (ion source), and allowed to diffuse to the ion source.

The two methods commonly used to produce ions from thermally volatile materials are electron impact (EI) and chemical ionization (CI). In the former technique, the

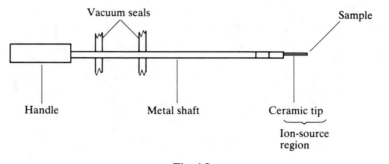

Fig. 4.2

probe tip shown in Fig. 4.2 is located near to a heated filament (Fig. 4.3). Electrons are accelerated from the hot filament to an anode, usually through a potential difference of about 70 V. Since 1 eV $\simeq$ 23 kcal mol^{-1} $\simeq$ 96 kJ mol^{-1}, a 70 eV electron has sufficient energy not only to ionize an organic molecule (requiring about 7–10 eV), but also to cause extensive fragmentation (the strongest single bonds in organic molecules have strengths of about 4 eV). In fact, a 70 eV electron by no means deposits all its energy into a molecule with which it interacts; typically, if it causes ionization, it will then deposit only 0–6 eV of internal energy in the resulting ion.

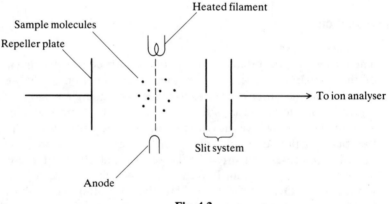

Fig. 4.3

Since organic molecules, almost without exception, contain electrons in pairs in filled orbitals, ionization of a sample molecule by removing an electron leaves behind an electron which is unpaired, i.e. the product is a cation-radical (**1**). Electron capture to give an anion-radical (**2**) does not occur to a significant extent since the bombarding electrons have such high translational energies that they cannot be captured.

$$M \xrightarrow{-e} M^{+\cdot} \qquad M \xrightarrow{+e} M^{-\cdot}$$

1   **2**

In the EI source shown in Fig. 4.3, the molecular ions produced in the electron beam can fragment either by loss of a radical, or by loss of a molecule with all its electrons paired. A process of each kind, occurring in the molecular ion of butyl acetate, is shown below, and the mass spectrum (Fig. 4.18) is discussed later.

$$CH_3COOCH_2CH_2CH_2CH_3^{+} \begin{array}{c} \nearrow CH_3\overset{+}{C}=O + C_4H_9O^{\cdot} \\ \\ \searrow CH_3COOH + C_4H_8^{+} \end{array}$$

The example illustrates both an advantage and a disadvantage of EI mass spectrometry. The advantage is that fragmentation is extensive, giving rise to a pattern of fragment ions which can help to characterize the compound [especially if it is a previously known compound, and the mass spectrum is already available for comparison in a computer file (Sec. 4.7)]. The disadvantage is the frequent absence of a molecular ion: in a typical selection of organic compounds, even in the limited molecular weight range up to 300 Daltons, 10–20 per cent may lack a molecular ion in their EI spectrum.

The mixture of molecular ions (if present) and fragment ions is expelled from the ionization chamber (Fig. 4.3) by a positive (repeller) voltage applied to the plate shown to the left of the ion source, and then goes into the ion analyser (Sec. 4.2).

The occasional lack of a molecular ion, or even the common production of a molecular ion of low abundance, in EI spectra is a serious disadvantage of the method. For example, if the supposed molecular ion (i.e. ion at highest m/z value) is only 1–2 per cent of the abundance of the most abundant fragment ions, how can we be confident that it is not in fact associated with a trace impurity? The problem is usually avoided by the use of CI mass spectrometry. In this technique a reagent gas (e.g. methane, isobutane, or ammonia) is allowed to pass into the ion chamber at a pressure of about 10^2 N m^{-2}. This gas is ionized by using electrons (produced from a hot filament as before) with energies up to 300 eV, e.g.

$$CH_4 + e \longrightarrow CH_4^{+} + 2e$$

Some fragmentation then occurs:

$$CH_4^{+} \longrightarrow CH_3^{+} + H^{\cdot}$$

However, in contrast to the situation occurring in EI mass spectrometry, the ion produced can now, at these higher source pressures, collide with its neutral counterparts. The main bimolecular reactions occurring are:

$$CH_4^{+} + CH_4 \longrightarrow CH_5^{+} + \dot{C}H_3$$

$$CH_3^{+} + CH_4 \longrightarrow C_2H_5^{+} + H_2$$

If sample ions are volatilized into this mixture of ions, the CH_5^{+} (which, for our purposes, can conveniently be regarded as CH_4 'solvating' H^+) acts as a strong acid and protonates the sample

$$M + CH_5^+ \longrightarrow MH^+ + CH_4$$

Thus in positive-ion CI spectra, molecular weight information is obtained from protonation of sample molecules, and the observed m/z value is one unit greater than the true molecular weight. In isobutane or ammonia CI spectra, the reagent ions (analogous to CH_5^+) causing protonation of sample are $C_4H_9^+$ and NH_4^+, respectively. The internal energy of MH^+, produced from CH_5^+, $C_4H_9^+$, or NH_4^+ decreases in the order of the reagent ions given. This is because M will typically bind a proton much more strongly than CH_5^+, and energy is therefore released in association with the proton transfer. Conversely, NH_4^+ binds a proton so strongly that it will not protonate some molecules (e.g. hydrocarbons); but if it does protonate the sample, the MH^+ ion produced will normally have so little internal energy that it will be very abundant in the resulting mass spectrum. Thus, ammonia CI is frequently an excellent technique for the determination of molecular weights of volatile molecules, but little fragmentation may be observed. Note that EI mass spectrometry may fail to give a clear indication of molecular weight not only because of the high internal energy of M^+, but also because M^+ may fragment extremely easily because of the instability associated with the unpaired electron. A schematic diagram of a CI source is given in Fig. 4.4.

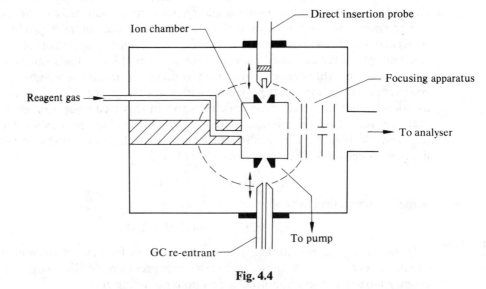

Fig. 4.4

(Reprinted with permission from Howe, Williams, and Bowen, 1981.)

Although EI does not produce satisfactory negative-ion spectra, negative-ion CI works well for molecules with electron-accepting properties (e.g. trifluoroacetates, quinones, and nitro-compounds). This is because the collisions occurring in a CI source reduce the large initial kinetic energies of the bombarding electrons to lower values at which they can be captured to give a radical-anion. Alternatively, a reagent ion such as CH_3O^- may be generated in the CI source. This can then act as a Bronsted base by abstracting a proton from the sample molecule.

$$M + CH_3O^- \longrightarrow (M-H)^- + CH_3OH$$

Involatile materials

The ionization methods discussed in this section should be used to obtain the mass spectra of molecules which may have low molecular weights (e.g. 200 Daltons) but have numerous polar functionalities, or have high molecular weights (800–10 000 Daltons).

Field desorption.   In this technique, the probe tip shown in Fig. 4.2 is replaced by a thin (e.g. 5 µm) wire on which sharp needles have been grown. The wire is supported between two posts on the probe (Fig. 4.5a).

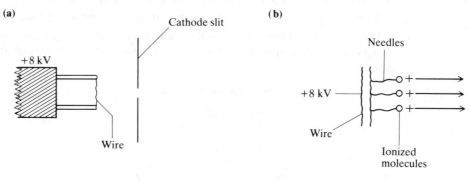

Fig. 4.5

A solution of a small amount of sample (e.g. 1 µg) is deposited on the wire. In the mass spectrometer, the wire is maintained at +8 kV, and can be heated. Since the needles on the wire are very sharp, the field at their tips may be as high as 10^8 V cm^{-1}, and this can cause the discharge of an electron from the sample into vacant orbitals of the metal from which the wire is made. Thus, positive ions (M^+) are created at the positive wire, and M^+ is largely desorbed by Coulombic repulsion (Fig. 4.5b), often with some help from limited heating of the wire. In this way, M^+ may be thrown into the gas phase without thermal decomposition. Additionally, the sample pressure at the end of a needle may be remarkably high due to the strong electric fields pertaining there. As a consequence, bimolecular reactions may occur, and field desorption (FD) spectra may on occasion show MH^+ to give molecular weight information, rather than M^+.

Desorption ionization by particles or radiation.   Between the late 1970s and the present time, four methods have been developed which allow in many cases the molecular weight determination of polar molecules in the mass range of about 300–25 000 Daltons. These are: laser desorption, fast atom bombardment (FAB), secondary ion mass spectrometry (SIMS), and californium plasma desorption. All these methods are based upon giving a large pulse of energy to the sample. The effect of this is to put a relatively large amount of energy into translational modes involving sample molecules. Thus, intermolecular bonds involving the sample—e.g. hydrogen bonds—are broken in preference to covalent bonds, and the sample is desorbed from its environment into the gas phase. Thermal decomposition is reduced, or avoided, by avoiding an equilibrium distribution of the large amount of available energy. This is because the sample molecule leaves its solid or liquid environment within a time of the order of 10^{-12} s.

In laser desorption, the energy is provided by a laser beam, whereas in SIMS or FAB it is provided by a beam of ions or atoms, respectively, of large translational energies (several keV). The sample may be bombarded in its solid state, but more commonly (and almost always in the case of FAB) is first dissolved in a matrix of low volatility. A schematic illustration of the FAB experiment is given in Fig. 4.6. Typically a few micrograms of sample are dissolved in a few microlitres of glycerol [$CH_2(OH)CH(OH)CH_2OH$] as matrix, and the solution is then bombarded by a beam of fast xenon atoms. These fast atoms ($\vec{Xe}$) are prepared by accelerating xenon ions ($\vec{Xe}^+$) to an energy in the range 6–9 keV, and then neutralizing these ions by charge exchange (electron transfer) as they pass almost stationary xenon atoms at low pressures.

$$\vec{Xe}^+ + Xe \longrightarrow \vec{Xe} + Xe^+$$

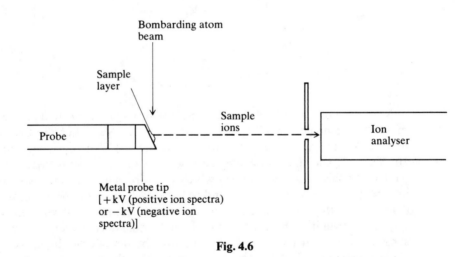

Fig. 4.6

When the fast xenon atoms impact into the solution of the sample in the matrix, the sample is desorbed, often as an ion, by momentum transfer. The beam of sample ions is then analysed in the mass spectrometer in the usual way. In view of the ability of this technique to analyse large polar molecules, it is beneficial to couple a FAB source with a magnet having a mass range up to 10 000–12 000 Daltons at full accelerating voltage. Often, protonated oligomers of the matrix [$(glycerol)_n H^+$, m/z

$(92n + 1)]$ are desorbed along with the sample ions, and may serve to mass mark the spectrum.

It is important that the sample should dissolve in the matrix, and preferably be marginally more hydrophobic than the matrix so that it will occupy the matrix/ 'vacuum' interface. This is desirable because the xenon beam only penetrates about 10 nm into the matrix. A commonly used alternative matrix to glycerol is thioglycerol/ diglycerol (1:1). More hydrophobic matrices which may on occasions be used to advantage are tetragol (3) and teracol (4).

$$HO(CH_2CH_2O)_4H \qquad HO(CH_2CH_2CH_2CH_2O)_nH$$
3   4

Positive- and negative-ion FAB mass spectra are recorded with similar facility, with molecular weights normally being given by abundant MH^+ ions in positive-ion spectra, and by abundant $(M—H^+)^-$ ions in negative-ion spectra. It is, of course, true that the neutral sample molecule M will also be desorbed, but since the polar molecules being analysed usually have relatively acidic (e.g. —COOH) or basic (e.g. —NH$_2$) sites, then the corresponding ions (—CO$_2^-$ or —NH$_3^+$) are desorbed in comparable amounts. FAB and SIMS spectra frequently contain structurally useful fragment ions, but in the spectra of large molecules obtained from a matrix, MH^+ is usually the most abundant ion, making molecular weight determination extremely easy by these techniques.

The principle of the californium (^{252}Cf) plasma desorption source is illustrated in Fig. 4.7. The sample to be analysed is deposited on a thin metal foil, usually of nickel. Spontaneous fission of the radioactive ^{252}Cf nucleus occurs, and each fission event gives rise to two fragments travelling in opposite directions (because of the necessity of momentum conservation). A typical pair of fission fragments is ^{142}Ba^{18+} and ^{106}Tc^{22+}, with kinetic energies of roughly 79 and 104 Mev, respectively. When such

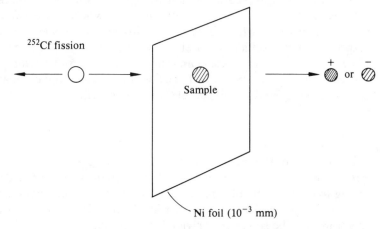

^{252}Cf fission

Sample

Ni foil (10^{-3} mm)

Fig. 4.7

(Reprinted with permission from Howe, Williams, and Bowen, 1981.)

a high-energy fission fragment passes through the sample foil, extremely rapid localized heating occurs, producing a 'temperature' in the region of 10 000 K. Consequently, the molecules in this plasma zone are desorbed, with the production of both positive and negative ions. These ions may then be accelerated out of the source into the analyser system.

^{252}Cf plasma desorption is better able to produce molecular ion signals than FAB or SIMS in the molecular weight range 10 000–20 000 Daltons, where the precision of the mass determination might typically be $\pm$10–20 mass units.

Sensitivity

Although all the methods for the production of ions which have been discussed are not ideal in terms of efficiency, the methods which are available for the detection of ions are so sensitive that mass spectrometry is an extremely sensitive technique. Mass spectra can be routinely obtained on a few micrograms of sample, and in the most favourable cases on picogram (10^{-12} g) samples. Hence, its enormous importance in solving problems where only a very small sample is available.

4.3   Ion analysis

Once the sample to be investigated has been volatilized and ionized, it is necessary to analyse the ions which are produced. In general, the ions are repelled out of the source using a small (<50 V) repeller voltage, transmitted through an accelerating potential (where required), and then injected into a mass analyser. The most important kinds of mass analyser are covered in the following subsections.

Magnetic-sector instruments

If we wish merely to separate all ions in the spectrometer which differ by at least unit mass (e.g. to resolve m/z 110 from m/z 111, where these values represent single-charged fragments whose atomic constituents add up to masses of 110 and 111, respectively), it is sufficient to deflect the ions in only a strong magnetic field. Ions of larger mass are deflected less than ions of smaller mass according to Eq. 4.1, where B is the strength of the magnetic field, r is the radius of the circular path in which the ion is travelling, and V is the accelerating potential.

$$\frac{m}{z} = \frac{B^2 r^2}{2V} \tag{4.1}$$

If the poles of the magnet are pictured as lying above and below the plane of the paper, then the radial paths followed by ions in the magnetic field are those illustrated in Fig. 4.8. It will be obvious that by scanning the magnetic field, Eq. 4.1 can be satisfied for ions of all m/z ratios for fixed values of r and V. Alternatively, the mass spectrum may be scanned electrically by varying V while the magnetic field is held constant. Whichever device is employed, ions of all m/z values can be successively allowed to pass through the collected slit D (Fig. 4.8) and the mass spectrum recorded.

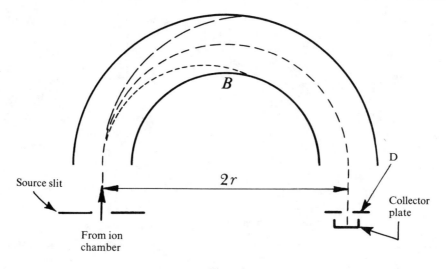

Source slit

From ion
chamber

$2r$

B

D

Collector
plate

Fig. 4.8

Frequently, we may wish to differentiate between ions which have the same nominal integral mass but possess different exact masses. This should be possible since, generally speaking, the isotopes of which elements consist do not have exact integral masses. Based on the convention that the atomic weight of ^{12}C is exactly 12, the masses of the most abundant isotopes of hydrogen, nitrogen, and oxygen are given in Table 4.1. Hence it will be evident that although CO, $CH_2{=}CH_2$, and N_2 all have the same integral mass (28), the exact masses of the four species are different, as indicated in the table.

Table 4.1 Exact masses of some common isotopes and simple molecular species, with the atomic mass of ^{12}C taken to be exactly 12

Species	^1H	^{16}O	^{14}N	CO	$CH_2{=}CH_2$	N_2
Mass	1.00782	15.9949	14.0031	27.9949	28.0313	28.0061

By using a high-resolution mass spectrometer, it is possible to separate the positive ions corresponding to CO, $CH_2{=}CH_2$, and N_2. High resolution is achieved by passing the ion beam through an electrostatic analyser before it enters the magnetic sector (Fig. 4.9). In such a double-focusing mass spectrometer, ion masses can be measured with an accuracy of about 1 p.p.m. With mass measurement of this accuracy, the atomic composition of the molecular ions can be determined, or the possibilities limited. However, precise mass measurement requires the use of narrow source exit and collector slits, with a consequent loss in sensitivity. The determination, or limitation of the possibilities, of the molecular formula of a sample is clearly a very important function of mass spectrometry.

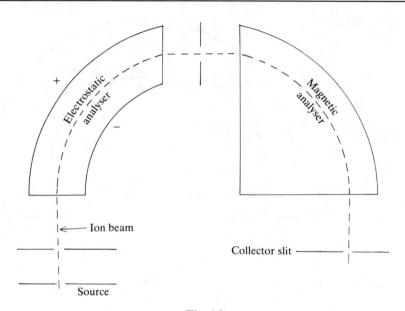

Fig. 4.9

In later sections, we will discuss the separation of mixtures by GC or LC prior to mass spectrometric analysis. However, both the functions of separation and analysis can be carried out by mass spectrometry alone. The simplest device to do this is a double-focusing mass spectrometer in which the electric sector follows the magnet sector (Fig. 4.10). If a mixture of three compounds of molecular weights M_1, M_2, and M_3 is ionized in the source, then the molecular ions of these can be separated in the magnetic analyser. The magnetic field is set to pass only M_2^+ through a slit placed

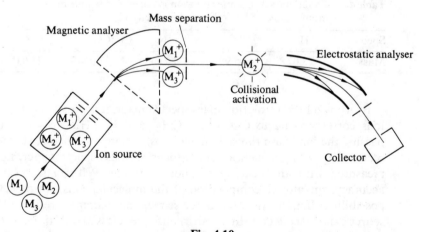

Fig. 4.10

(Reprinted with permission from Howe, Williams, and Bowen, 1981.)

between the magnetic and electric sectors. A collision chamber is located behind this slit. This chamber contains an inert gas at 10^{-3}–10^{-2} N m^{-2}, so that when M_2^+ has a grazing collision with an inert gas atom, a tiny proportion of its translational (kinetic) energy is converted into internal energy of vibration. As a result, fragment ions are produced from M_2^+. Suppose a fragment ion retains x per cent of the mass of M_2^+, i.e. $(100-x)$ per cent of the mass is lost as a neutral particle. This means that the fragment retains only x per cent of the translational energy of M_2^+, the translational energy being partitioned in the ratio of the product masses. Since the electrostatic analyser deflects ions of lower kinetic energy more, it separates the ions produced by decomposition of M_2^+ according to their kinetic energies, and hence in this case according to their masses. By scanning the electric sector voltage downwards from an initial value E_0 (which allows M_2^+ to pass through the collector slit), the products of the collisionally induced decomposition are detected at the collector in order of decreasing mass.

The above experiment has produced a collisionally induced mass spectrum of M_2^+, and these spectra are similar to the mass spectra obtained when the energy is deposited by other means. Since all components of a mixture M_1, M_2, ..., M_n can be analysed in this way, all components of a mixture (other than isomers) can in principle be separated, and a mass spectrum obtained for each one. The technique is commonly called MS/MS, since the molecular ion from an initial mass spectrum is selected and made to give a second mass spectrum. The potential power and elegance of the method are obvious, but there are limitations:

1. The molecular ions produced from the mixture must be abundant, in order to give good sensitivity.
2. The fragmentation of the molecular ions in the ion source should be limited, to avoid fragment ions formed there having, by chance, the same mass as molecular ions from other components of the mixture.
3. The mass resolution of the electrostatic analyser in the above arrangement (Fig. 4.10) is limited to a few hundred mass units.
4. If the mixture contains substances whose structures are novel (i.e. not simply at present unknown to the investigator, but ones which have not previously been determined), then one mass spectrum will not normally be sufficient to determine the structure (peptides may be an exception). Therefore, the structure determination of novel substances (peptides excepted) will in any event require HPLC or GLC (or some other) separation of the components of the mixture. In contrast, if collision-induced mass spectra of the components of the mixture are already available and in a computer file, MS/MS can be very powerful.

The solution to problems 1 and 2 is to use FAB or SIMS, or possibly CI, as the ionization technique. Problem 3 can be solved by doing MS/MS with two mass spectrometers in tandem. If an MS/MS experiment is performed by separating the molecular ions of a mixture in one double-focusing mass spectrometer, then causing these ions to decompose by collision-induced decomposition, and finally separating these collision-induced products in a second high-resolution mass spectrometer (Fig. 4.11), then problem 3 above is removed.

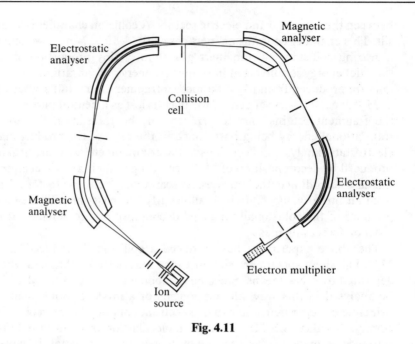

Fig. 4.11

Quadrupole mass spectrometers

Quadrupole mass spectrometers are commercial rivals of magnetic sector instruments, especially in the range up to m/z 1000, where high resolution is not required, and where simplicity of operation may be an advantage. They are used particularly where GC and/or HPLC instruments are directly coupled to the mass spectrometer, although magnetic sector instruments also perform this function well.

The arrangement of electrodes in a quadrupole mass filter is given in Fig. 4.12. A constant voltage U and a radiofrequency potential V are applied between opposite pairs of four parallel rods. The rods are between 0.1 and 0.3 m long in most commercial instruments.

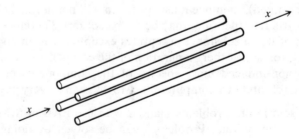

Fig. 4.12

Ions are injected along the x-direction, and the mass spectrum scanned either by varying the amplitude of U and V, while keeping the ratio U/V constant, or by varying the frequency of the radiofrequency potential.

Mixture analysis can be achieved by three quadrupole mass spectrometers connected in sequence. The first quadrupole is used to separate the molecular ions, the second as a collision chamber, and the third to separate the products of collision-induced decomposition (cf. Fig. 4.10).

Time-of-flight mass spectrometers

All singly charged ions dropping through a potential difference V will acquire the same translational energy eV. Therefore, those of largest mass will have the lowest velocities, and the longest time of flight over a given distance. This property is used in the time-of-flight mass spectrometer (Fig. 4.13), in which a voltage pulse on grid A extracts the ions from the source. The ions are then accelerated by a potential difference between A and B, and subsequently pass into the field-free flight tube. They are separated in time, according to their m/z values, and collected at D. Since it is common to have differences in arrival times between successive mass peaks of $\leqslant 10^{-7}$ s, fast electronics are required to distinguish successive peaks. This method of ion analysis is used in conjunction with ^{252}Cf desorption (Sec. 4.2) since the fission fragment complementary to that causing ionization can be used as a zero-time marker.

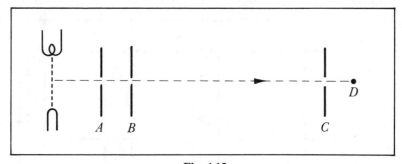

Fig. 4.13

(Reprinted with permission from Howe, Williams, and Bowen, 1981.)

Ion-cyclotron resonance (ICR) spectrometers

In this method of ion analysis, the ions to be analysed are injected into an ICR cell at low translational energies. Within the cell, a uniform magnetic field B constrains the ions to a circular path perpendicular to the direction of the magnetic field. For a singly charged ion, their frequency (ω_c) is given by:

$$\omega_c = eB/m$$

If an alternating electric field of frequency ω_1 is applied normal to B, an ion will absorb energy if $\omega_c = \omega_1$. Thus, an ICR mass spectrum can be obtained by fixing B and scanning ω_1 so that ions of different mass successively satisfy the above equation. Absorption of energy by the ions (at resonance) is measured using an oscillator detector system similar to that used in NMR (Chapter 3). By further analogy to NMR, a spread of frequencies may be generated by using a pulsed (radiofrequency) electric field, and the spectrum obtained by Fourier transform methods (Secs 2.3 and 3.1). For good sensitivity, the ions should be kept resonating for times in the range of milliseconds–seconds. To achieve this, very low operating pressures (about 10^{-6} N m^{-2}) are required. If these can be achieved, the method has the advantage of very high sensitivity and high resolving power.

4.4 Ion abundances in mass spectra

Isotope abundances

All singly charged ions in the mass spectrum which contain carbon also give rise to a peak at one mass unit higher. This happens because of the natural abundance of ^{13}C (1.1 per cent). For an ion containing n carbon atoms, the abundance of the isotope peak is $n \times 1.1$ per cent of the ^{12}C-containing peak. Thus $C_5H_{12}^+$, $C_{40}H_{70}^+$, and $C_{100}H_{170}^+$ would give isotope peaks at one mass unit higher of approximate abundances 5.5, 44, and 110 per cent of the abundance of the ions containing ^{12}C only. Obviously, the probability of finding two ^{13}C atoms in an ion is very low, and M + 2 peaks are accordingly of low abundance (although this is not true for very large molecules).

Although iodine and fluorine are monoisotopic, chlorine consists of ^{35}Cl and ^{37}Cl in the ratio of approximately 3:1 and bromine of ^{79}Br and ^{81}Br in the ratio of

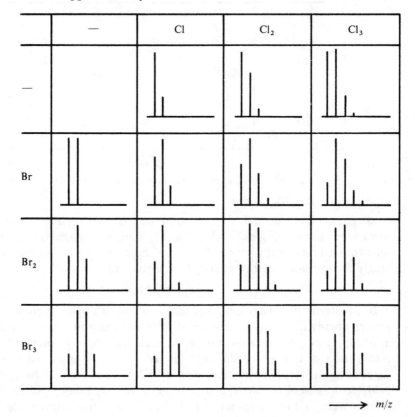

$\longrightarrow$ m/z

Fig. 4.14

approximately $1:1$. Molecular ions (or fragment ions) containing various numbers of chlorine and/or bromine atoms therefore give rise to the patterns shown in Fig. 4.14 (all peaks spaced 2 mass units apart).

Obviously, the isotope patterns to be expected from any combination of elements can readily be calculated, and provide a useful test of ion composition in those cases where polyisotopic elements are involved. Most of the remaining elements in Table 4.2 are essentially monoisotopic, with the exception of sulphur and silicon.

Table 4.2   Atomic weights and approximate natural abundance of some isotopes

Isotope	Atomic weight ($^{12}C = 12.000\,000$)	Natural abundance (%)
^1H	1.007 825	99.985
^2H	2.014 102	0.015
^{12}C	12.000 000	98.9
^{13}C	13.003 354	1.1
^{14}N	14.003 074	99.64
^{15}N	15.000 108	0.36
^{16}O	15.994 915	99.8
^{17}O	16.999 133	0.04
^{18}O	17.999 160	0.2
^{19}F	18.998 405	100
^{28}Si	27.976 927	92.2
^{29}Si	28.976 491	4.7
^{30}Si	29.973 761	3.1
^{31}P	30.973 763	100
^{32}S	31.972 074	95.0
^{33}S	32.971 461	0.76
^{34}S	33.967 865	4.2
^{35}Cl	34.968 855	75.8
^{37}Cl	36.965 896	24.2
^{79}Br	78.918 348	50.5
^{81}Br	80.916 344	49.5
^{127}I	126.904 352	100

Energetics of fragmentation

It is true that the longer we keep an ion before detecting it, the more chance it has to decompose to a daughter ion. However, as a crude (but for our purposes simplifying and helpful) approximation, we can ignore this effect. The reason we can ignore it is that once the activation energy is available for a unimolecular decomposition, the reaction mainly occurs on a time scale which is shorter than the ion lifetime before detection.

The ionization methods discussed in Sec. 4.2 invariably produce molecular ions with varying internal energies E. The probability $[p(E)]$ of any given ion having a specific value of E is not known, but it is sufficient to know that a range of energies will exist which will be lower for the so-called soft-ionization methods (e.g. CI, FD, FAB, SIMS from a matrix, ^{252}Cf) than for EI. Suppose the distribution of internal energies is roughly of the form shown in Fig. 4.15, and E_0 is the activation energy for the most favoured decomposition of M^+ (or MH^+) to A^+. Then roughly speaking, those M^+ ions which possess energies less than E_0 will be recorded as M^+. Conversely, those possessing an energy greater than E_0 will be recorded as A^+ [or its decomposition products, since an identical analysis could be presented for each decomposition step (e.g. $A^+ \rightarrow B^+$, etc.)].

Although the above approximation is a good and useful one, it will be clear that a few molecular ions will have energies so little in excess of E_0 that they will not decompose within the $\sim 10^{-5}$ s needed for the ion to travel from source to collector, and hence will still be recorded as molecular ions. Still a few other molecular ions with internal energies only slightly in excess of E_0 will actually survive long enough to be accelerated (through an accelerating potential V, as they leave the source) but decompose to a daughter ion (A^+) in the field-free region between the source exit slit and the beginning of the magnetic analyser (see Fig. 4.8). If we consider not just the special case $M^+ \rightarrow A^+$, but any ion decomposition $m_1 \rightarrow m_2$, then the product ion m_2 no longer possesses the normal translational energy zV of a parent or daughter ion formed in the source (due to an ion of charge z falling through an accelerating voltage V). Rather, the translational energy zV of m_1 is partitioned between m_2 and the neutral particle (m_1-m_2) in the ratio of their masses. Hence, the translational energy of m_2 is $m_2 zV/m_1$, lower by a factor m_2/m_1 than that of a normal m_2 formed in the source. The m_2 ion formed in the field-free region does not therefore appear at m_2 on the mass scale but at a lower value m^* given by the relationship:

$$m^* = \frac{m_2}{m_1}m_2 = \frac{m_2{}^2}{m_1}$$

Such peaks are known as metastable peaks, and normally their presence suggests that the reaction $m_1 \rightarrow m_2$ occurs in one step. Metastable peaks are recognizable because they can occur at non-integral values, and are broader and much less abundant than normal peaks. If metastable peaks are observed in a mass spectrum,

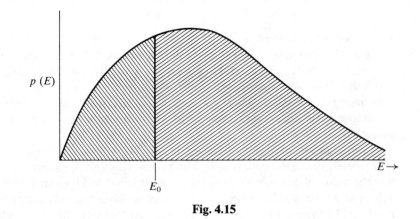

Fig. 4.15

then the $m_1 \rightarrow m_2$ transition associated with them can usually be worked out using the above equation. The ability to relate from one mass peak (m_1) to another (m_2) in a mass spectrum in this way is often useful in deducing structural features from the mass spectrum of an unknown. For this reason, the masses (m^*) of metastable peaks occurring in many of the mass spectra reproduced in Chapter 5 are given with the spectra.

The value of the approximation given in Fig. 4.15, whereby the relative M^+ and A^+ abundances are given by the hatched areas to the left and right of E_0, respectively, is several-fold. First, it shows us that if a molecular ion can decompose to two fragments without an activation energy, the molecular ion will be absent in the mass spectrum. Conversely, a decomposition requiring $\geqslant 4\,\text{eV}$ in activation energy will give (by any of the ionization techniques discussed) an abundant molecular ion and a relatively low abundance of fragment ions.

As a further approximation, we can say that fragmentations corresponding to single-bond cleavages will occur with little or no reverse activation energy (Fig. 4.16a). In contrast, fragmentation involving loss of stable molecules may involve reverse activation energies (Fig. 4.16b).

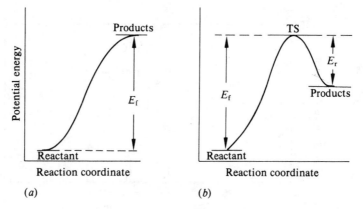

Fig. 4.16

Thus, for a single-bond cleavage, the activation energy for fragmentation will be approximated by:

$$[\Delta H_f(\text{products}) - \Delta H_f(\text{reactant})]\ \text{kJ mol}^{-1}$$

The principle is illustrated by reference to the 70 eV mass spectrum of ethylbenzene. Clearly, the bonds of the benzene ring will be stronger than those involved in the sidechain, and so it seems sensible to search for the most energetically favourable unimolecular fragmentation of the molecular ion from the three following possibilities.

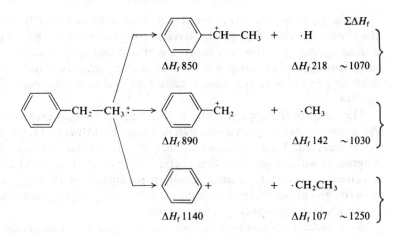

It is clear from the ΔH_f data that loss of a methyl radical should be favoured over the other possibilities, as is observed in the mass spectrum (Fig. 4.17). Note that since sp^2 hybridized carbonium ions (and in particular the $C_6H_5{}^+$ cation) have high ΔH_f values, their formation is in general unfavourable. Exceptions arise when the radicals lost in their formation have particularly low ΔH_f values, and/or there is no alternative

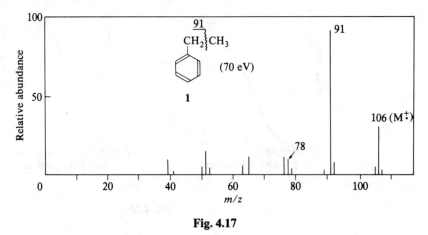

Fig. 4.17

(Reprinted with permission from Howe, Williams, and Bowen, 1981.)

lower energy fragmentation, e.g. in the mass spectra of C_6H_5I or C_6H_5Br. Additionally, it can be seen from the above data that the secondary benzylic carbonium ion has a lower ΔH_f value (850 kJ mol^{-1}) than the primary benzyl cation (890 kJ mol^{-1}), but that this effect in the present case is more than offset by the greater stability of a methyl radical relative to a hydrogen radical.

The activation energy for formation of m/z 91 is approximately 160 kJ mol^{-1} ($1030 - 870$, the latter number being ΔH_f for ionized ethylbenzene), and therefore the molecular ion in Fig. 4.17 is of moderate abundance.

Some routes which might be considered for the decomposition of ionized butyl acetate in its EI spectrum are given in Table 4.3.

Table 4.3   Some hypothetical decomposition routes for $[BuOAc]^+$ (Reprinted with permission from Howe, Williams, and Bowen, 1981.)

Reaction	Products	m/z	$\Sigma \Delta H_f, kJ\ mol^{-1}$
1	$CH_3\overset{+}{C}{=}O + \cdot OCH_2CH_2CH_2CH_3$	43	~560
2	$CH_3\overset{O}{\overset{\|}{C}}{-}\overset{+}{O}{=}CH_2 + \cdot CH_2CH_2CH_3$	73	~530
3	$CH_3\overset{O}{\overset{\|}{C}}OCH_2CH_2 \cdot + \overset{+}{C}H_2CH_3$	29	~770
4	$CH_3\overset{O}{\overset{\|}{C}}{-}\overset{+}{O}{=}CHCH_3 + \cdot CH_2CH_3$	87	~510
5	$CH_3COOH^+ + CH_2{=}CHCH_2CH_3$	60	580
6	$CH_2{=}CHCH_2CH_3{}^+ + CH_3COOH$	56	450
7	$CH_3C\overset{\overset{\displaystyle +OH}{\diagup\!\diagup}}{\diagdown_{\displaystyle OH}} + CH_2{=}CH\dot{C}HCH_3$	61	470
8	$CH_3CH_2CH_2CH_2CH_3{}^+ + CO_2$	72	460
9	$CH_3OCH_2CH_2CH_2CH_3{}^+ + CO$	88	~530

$\Delta H_f[BuOAc]^+$ is approximately 430 kJ mol^{-1}, and the thermal energy alone of a molecule of the size of butyl acetate at 100°C is of the order of 100 kJ mol^{-1}. Therefore, if a reaction such as (6) (which gives rise to a particularly favourable combination of products, Table 4.3) occurs with a small or negligible reverse activation energy, then the abundance of M^+ should be small or negligible. The hypothetical reactions of Table 4.3 are split into three groups: single-bond cleavages, (1)–(3); fragmentations with hydrogen rearrangements, (4)–(7); and skeletal reorganization with extrusion of extremely stable neutrals, (8) and (9). Clearly among many possible single-bond cleavage reactions, (1) and (2) should be the most favourable ones, as is observed (Fig. 4.18a). The spectrum also shows that single [reaction (6)] and even double [reaction (7)] hydrogen rearrangements are observed in mass spectra where they are energetically (Table 4.3) and geometrically (Scheme 4.1) favourable. In the scheme, the molecular ion is depicted as being formed by removal of a lone-pair electron from oxygen (since such an electron has a low binding energy). The unpaired electron which remains is then used to promote the hydrogen rearrangement. Ion-dipole forces can transiently hold together the initial products (indicated in square brackets) with subsequent dissociation, accompanied either by electron or hydrogen radical transfer, to stable combinations of products.

Scheme 4.1

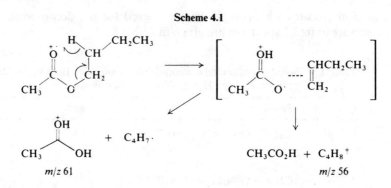

Despite the fact that both the hypothetical reactions (8) and (9) give rise to low energy products, they are not seen in practice (Fig. 4.18a). This observation is a useful general guide, i.e. reactions requiring σ-bond formation (other than to H) occur only exceptionally; the reverse activation energies are evidently too large. When the

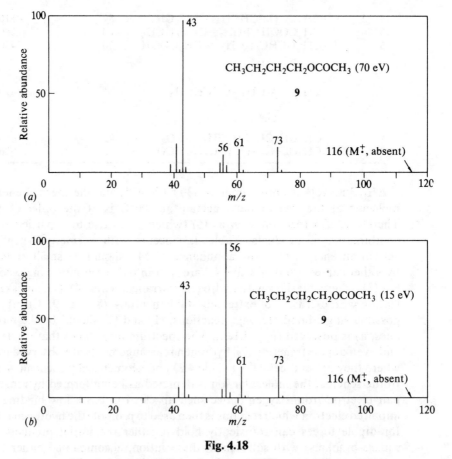

Fig. 4.18

(Reprinted with permission from Howe, Williams, and Bowen, 1981.)

exceptions occur it is usually found that σ-bond formation between initially separated parts of a molecule A and C, with loss of a stable neutral B, only occurs if A and/or C contain sites of unsaturation. If these cases of skeletal rearrangement were not limited, then mass spectrometry would be more limited as an aid to structure elucidation.

$$[ABC]^+ \longrightarrow [AC]^+ + B$$

The lowering of the electron beam energy to approximately 15 eV (Fig. 4.18b) does not (and should not) result in the observation of a molecular ion. CI should be used in this case for molecular weight determination. The lowering of the electron beam energy does, however, in general promote hydrogen rearrangement reactions at the expense of single-bond cleavages. This is because hydrogen rearrangement reactions require relatively specific geometries, and therefore their fastest rates are much slower than single-bond cleavages (which can occur in essentially any geometry). This point emphasizes that the appearance of a mass spectrum depends very much on the conditions under which it was obtained.

The above principles are the main ones which control the appearance of *all* mass spectra. However, it is not normally possible to interpret mass spectra by using large tables of ΔH_f values not only because insufficient data are available, but also because the ion or neutral structure is often unknown. Yet is is necessary to know whether a suggested combination of fragments is reasonable or not, and therefore some representative values ΔH_f are given in Tables 4.4 and 4.5.

Table 4.4 ΔH_f (kJ mol^{-1}) of some ions

Ion	ΔH_f	Ion	ΔH_f
H^+	1530	$(CH_3)_2\overset{+}{C}-CH=CH_2$	770
$CH_3{}^+$	1090	$C_6H_5{}^+$	1140
$C_2H_5{}^+$	920	$C_6H_5CH_2{}^+$	890
$CH_3CH_2CH_2CH_2{}^+$	840	$CH_3CH_2\overset{+}{C}=O$	600
$CH_3CH_2\overset{+}{C}HCH_3$	770	$C_6H_5\overset{+}{C}=O$	730
$(CH_3)_3C^+$	690	$CH_3CH=\overset{+}{O}H$	600
$CH_2=CH^+$	1110	$CH_3\overset{+}{O}=CH_2$	640
$\overset{+}{C}H_2-CH=CH_2$	950	$CH_3CH=\overset{+}{N}H_2$	650

Points and trends to note from Table 4.4 are:

1. Formation of very small ions such as H^+ and $CH_3{}^+$ is extremely unfavourable.
2. Vinyl cations have high ΔH_f values.
3. The order of ease of formation of cations is tertiary > secondary > primary.
4. Ions whose formation is favoured in solution, either as stable species or transient intermediates, are also relatively stable in the gas phase (e.g. delocalized cations, and acylium, oxonium, or imminium ions).

Table 4.5 ΔH_f (kJ mol^{-1}) of some radicals

Radical	ΔH_f	Radical	ΔH_f
H·	218	$C_6H_5CH_2$·	188
CH_3·	142	·OH	39
C_2H_5·	107	·OCH_3	−4
$CH_3CH_2CH_2$·	87	$CH_3\dot{C}{=}O$	−23
$CH_3\dot{C}HCH_3$	74	·NH_2	172
$CH_2{=}\dot{C}H$	250	·Cl	122
$CH_2{=}CH{-}\dot{C}H_2$	170	·Br	112
C_6H_5·	300	·I	107

Note the relative instability of vinyl radicals, and the increased stability of radicals with increasing substitution.

Some guides to interpretation, and examples of spectra obtained by various ionization methods follow.

Recognition of molecular ions
The apparent loss of 14 Daltons from a supposed molecular ion suggests the presence of a homologue (because :CH_2 has a very high ΔH_f value).

If a peak occurs 4–13 mass units below a supposed molecular ion, then it is probable that either the higher mass peak is not a molecular ion, or the spectrum is of a mixture of compounds.

The molecular weights of compounds containing only C, H, (O) are even, as are those of molecules additionally containing an even number of nitrogen atoms. The molecular weights of compounds containing only C, H, N(O) are odd when the number of nitrogens is odd.

EI spectra
Some generalizations on molecular ion abundances in EI spectra are given in Table 4.6.

Table 4.6 Molecular ion abundances in relation to molecular structure

Strong	Medium†	Weak or absent
	Conjugated olefins	
Aromatic hydrocarbons (ArH)	Ar⫢Br	Long chain aliphatic compounds
ArF	Ar⫢I	Branched alkanes
ArCl	ArCO⫢R	Tertiary aliphatic alcohols
ArCN	$ArCH_2$⫢R	Tertiary aliphatic bromides
$ArNH_2$	$ArCH_2$⫢Cl	Tertiary aliphatic iodides

† In this column, wavy lines indicate a relatively weak bond.

Aromatic compounds. Molecular ions from compounds of the general formula C_6H_5X will fragment by loss of X, or part of X. The energies required for these processes increases from the top left entry of Table 4.7 to that on the bottom right. One of these molecular ion fragmentations is seen in Fig. 4.19.

Table 4.7 Order of 'ease of fragmentation' of some C_6H_5X compounds

X	Neutral fragments lost from M^+	X	Neutral fragments lost from M^+
$COCH_3$	CH_3	OH	CO
$C(CH_3)_3$	CH_3	CH_3	H
$CH(CH_3)_2$	CH_3	Br	Br
CO_2CH_3	OCH_3	NO_2	NO_2, NO
$N(CH_3)_2$	H	NH_2	HCN
CHO	H	Cl	Cl
C_2H_5	CH_3	CN	HCN
OCH_3	CH_2O, CH_3	F	C_2H_2, HF
I	I	H	C_2H_2

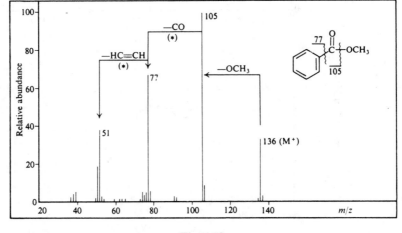

Fig. 4.19

In the mass spectra of disubstituted benzenes (within the limitations given below), the energetically easier fragmentation (Table 4.7) will be the one observed. Thus *p*-cyano-*t*-butylbenzene loses only a methyl radical as a primary reaction, and HCN loss from M^+ does not occur (Fig. 4.20a). A bromine radical is lost exclusively from the molecular ion of *p*-bromoaniline (Fig. 4.20b), but the table is of limited use when substituent groups fall very close together. For example, the molecular ion of *p*-chloroaniline competitively loses Cl and HCN (Fig. 4.20c), rather than losing exclusively HCN.

The secondary fragmentation of m/z 144 in Fig. 4.20a involving loss of ethylene might initially seem surprising, since m/z 144 will be produced as **5**. However, the energy of activation for the carbonium ion isomerization **5 → 6** is less than that for unimolecular decomposition of **5**. Ethylene is then lost from the reorganized carbon skeleton of **6**.

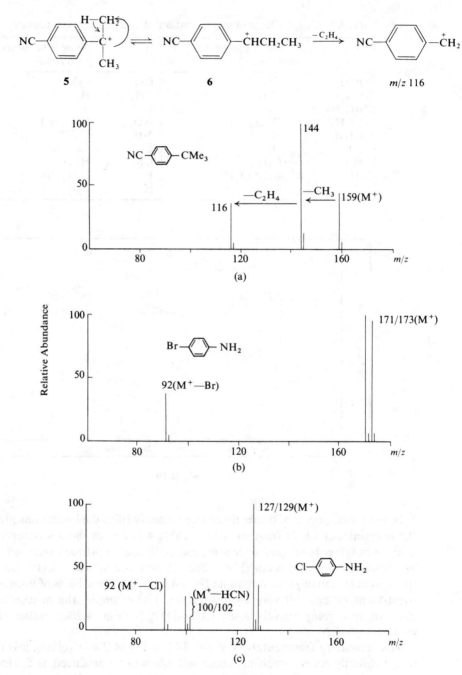

Fig. 4.20

The generalization drawn from Table 4.7 for the spectra of disubstituted benzenes can usefully be extended to *all* aromatic compounds (including heterocycles), and to greater degrees of substitution, with the following reservations.

Resonance effects may promote (**7 → 8**), or discriminate against (**9 → 10**) certain fragmentations,

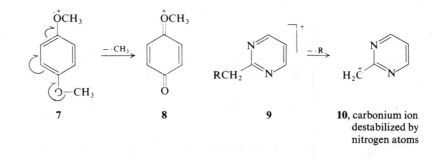

7	**8**	**9**	**10**, carbonium ion destabilized by nitrogen atoms

In *ortho*-substituted compounds, the spatial proximity of two groups may lead to new fragmentation pathways (e.g. **11 → 12** and **13 → 14**).

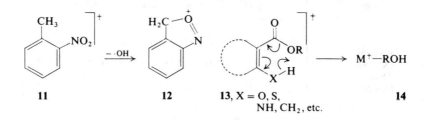

11	**12**	**13**, X = O, S, NH, CH$_2$, etc.	**14**

Aliphatic compounds. The cleavages associated with some common functionalities are given in Table 4.8; on passing down the table, the energy requirements for the processes increase. When primary fragmentation can lead to competition between losses of two or more different radicals, in 70 eV spectra the loss of the larger radical is usually dominant.

The m/z values associated with some common ion series are given in Table 4.9. If the groups attached to these 'simplest ion types' are saturated hydrocarbon groups, than the m/z values fall in the series m/z $(x + 14n)$, where $(m/z)x$ is the first member of the series (Table 4.9).

The elimination of neutral molecules directly from the molecular ions of aliphatic secondary and tertiary amines, ketals, iodides, and ethers is not normally observed; the primary processes of radical loss (Table 4.8) are too favourable. However, in nearly all aliphatic carbonyl compounds, a primary rearrangement fragmentation is observed, in which a γ-hydrogen, if available, migrates to the carbonyl oxygen. Usually, a charged enol is formed with elimination of a neutral olefin. The m/z values of the ions formed from various carbonyl compounds are given in Table 4.10.

Table 4.8 Primary single-bond cleavage processes associated with some common functional groups†

Functional group	Fragmentation
Amine	$R_2(CH_2)_n\overset{+}{\underset{R_1}{N}}{-}CH_2{-}R_3 \xrightarrow{-R_3\cdot} R_2(CH_2)_n\overset{+}{\underset{R_1}{N}}{=}CH_2 \xrightarrow{-\text{olefin}} H\overset{+}{\underset{R_1}{N}}{=}CH_2$
Ketal	$\xrightarrow{-R_2\cdot}$
Iodide	$R{-}I \rceil^+ \xrightarrow{-I\cdot} R^+$
Ether (X = 2O) Thioether (X = 2S)	$R_2(CH_2)_n\overset{+}{\underset{R_3}{X}}{-}CH{-}R_1 \xrightarrow{-R_3\cdot} R_2(CH_2)_n\overset{+}{X}{=}CH{-}R_1 \xrightarrow{-\text{olefin}} H\overset{+}{X}{=}CHR_1$
Ketone	$\underset{R_2}{\overset{R_1}{>}}C{=}\overset{+}{O} \xrightarrow{-R_2\cdot} R_1{-}C{\equiv}\overset{+}{O} \xrightarrow{-CO} R_1^+$
Alcohol (X = 2O) Thiol (X = S)	$R_1{-}CH_2{-}\overset{\cdot+}{\underset{R_2}{X}}H \xrightarrow{-R_2\cdot} R_1CH{=}\overset{+}{X}H$
Bromide	$RBr \rceil^+ \xrightarrow{-Br\cdot} R^+$
Ester	$R_1\overset{\cdot+}{\overset{\|}{\underset{}{C}}}OR_2 \longrightarrow \overset{+}{O}{\equiv}C{-}OR_2$ $\qquad \longrightarrow R_1C{\equiv}O^+$

† In polyfunctional aliphatic molecules, cleavages associated with functional groups higher up the table are preferred over those cleavages associated with lower entries.

Table 4.9   Useful ion series

Functional group	Simplest ion type	Ion series (m/z)
Amine	$CH_2{=}\overset{+}{N}H_2$ m/z 30	30, 44, 58, 72, 86, 100, ...
Ether Alcohol	$CH_2{=}\overset{+}{O}H$ m/z 31	31, 45, 59, 73, 87, 101, ...
Ketone	$CH_3C{\equiv}\overset{+}{O}$ m/z 43	43, 57, 71, 85, 99, 113, ...
[Hydrocarbon]	$C_2H_5{}^{+}$ m/z 29	29, 43, 57, 71, 85, 99, 113, ...

Table 4.10 m/z values of some rearrangement ions found in the mass spectra of carbonyl compounds

Compound	X	m/z
Aldehyde	H	44
Ketone (methyl)	CH_3	58
Ketone (ethyl)	C_2H_5	72
Acid	OH	60
Ester (methyl)	OCH_3	74
Ester (ethyl)	OC_2H_5	88
Amide	NH_2	59

Examples of spectra of aliphatic compounds

Ketones

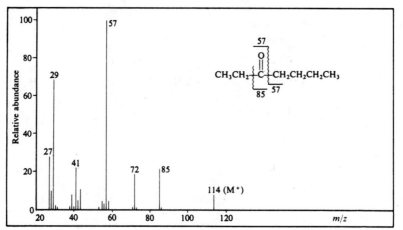

Fig. 4.21

1. Loss of alkyl groups attached to the carbonyl function (Table 4.8) leads to acylium ions $C_4H_9C\equiv O^+$ (Table 4.9).
2. Rearrangement ion at m/z 72 suggests ethyl ketone (Table 4.10).
3. m/z 57 shown by high resolution to be due to both $C_4H_9^+$ and $C_2H_5C\equiv O^+$ [saturated carbonium and acyl ions give same m/z values (Table 4.9)].

Amines

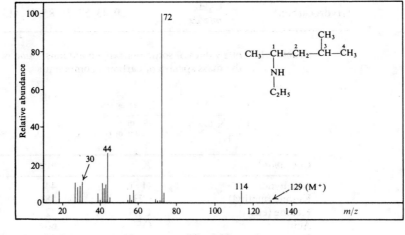

Fig. 4.22

1. Odd molecular weight and ions at m/z 30, 44, 72, and 114 suggest a saturated amine (Table 4.9).
2. Largest alkyl radical preferentially lost in decomposition of M^+:

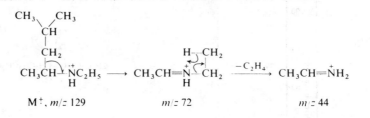

Two bifunctional molecules

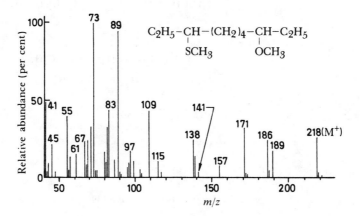

Fig. 4.23

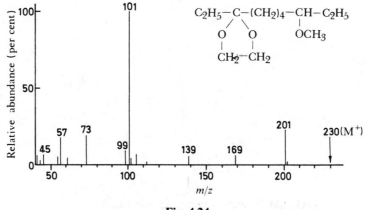

Fig. 4.24

1. The formation of sulphonium (m/z 89) and oxonium (m/z 73) ions is competitive in Fig. 4.23. However, in Fig. 4.24, the ion m/z 101 is so stable that pathways to other ions compete relatively poorly (cf. Table 4.8).
2. Decomposition pathways are summarized in Schemes 4.2 and 4.3.

The spectra of all these aliphatic compounds point to the fact that the molecular weight determination is better made by CI. A major use of EI is, therefore, to provide a sensitive fingerprint of a compound, particularly in GC/MS (Sec. 4.5). Since compounds frequently undergo extensive handling and/or chromatography prior to EI mass spectrometry, Table 4.11 gives a list of some peaks due to common impurities.

Scheme 4.2

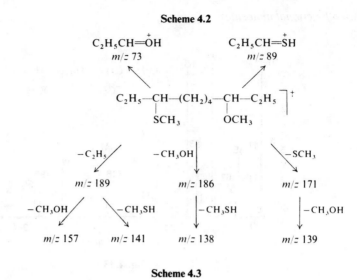

Scheme 4.3

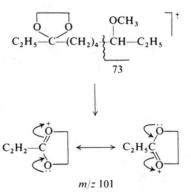

Table 4.11   Some common impurity peaks

m/z values	Cause
149, 167, 279	Plasticizers (phthalic acid derivatives)
129, 185, 259, 329	Plasticizer (tri-n-butyl acetyl citrate)
133, 207, 281, 355, 429	Silicone grease
99, 155, 211	Plasticizer (tributyl phosphate)

Once the molecular ion in an EI spectrum has been identified with reasonable certainty, in the case of other than very simple small molecules, it is usually advantageous to search for neutrals which are lost from M^+ (rather than attempt to identify a fragment ion from its m/z value). To facilitate such a search, Table 4.12 has been compiled.

Table 4.12   Some common losses from molecular ions

Ion	Groups commonly associated with the mass lost	Possible inference
M − 1	H	—
M − 2	H_2	—
. . . .	. . . .	. . . .
M − 14	—	Homologue?
M − 15	CH_3	—
M − 16	O	Ar—NO_2, $\geqslant \overset{+}{N}—\overset{-}{O}$, sulphoxide
M − 16	NH_2	$ArSO_2NH_2$, —$CONH_2$
M − 17	OH	—
M − 17	NH_3	—
M − 18	H_2O	Alcohol, aldehyde, ketone, etc.
M − 19	F	⎱ Fluorides
M − 20	HF	⎰
. . . .	. . . .	. . . .
M − 26	C_2H_2	Aromatic hydrocarbon
M − 27	HCN	⎧ Aromatic nitriles ⎩ Nitrogen heterocycles
M − 28	CO	Quinones
M − 28	C_2H_4	⎧ Aromatic ethyl ethers ⎩ Ethyl esters, n-propyl ketones
M − 29	CHO	—
M − 29	C_2H_5	Ethyl ketones, Ar—n-C_3H_7
M − 30	C_2H_6	—
M − 30	CH_2O	Aromatic methyl ether
M − 30	NO	Ar—NO_2
M − 31	OCH_3	Methyl ester
M − 32	CH_3OH	Methyl ester
M − 32	S	—
M − 33	$H_2O + CH_3$	—
M − 33	HS	⎱ Thiols
M − 34	H_2S	⎰
. . . .	. . . .	. . . .
M − 41	C_3H_5	Propyl ester
M − 42	CH_2CO	⎧ Methyl ketone ⎩ Aromatic acetate, $ArNHCOCH_3$
M − 42	C_3H_6	⎧ n- or iso-butyl ketone, ⎩ Aromatic propyl ether, Ar—n-C_4H_9
M − 43	C_3H_7	Propyl ketone, Ar—n-C_3H_7
M − 43	CH_3CO	Methyl ketone
M − 44	CO_2	⎧ Ester (skel. rearr.) ⎩ Anhydride
M − 44	C_3H_8	—
M − 45	CO_2H	Carboxylic acid
M − 45	OC_2H_5	Ethyl ester
M − 46	C_2H_5OH	Ethyl ester
M − 46	NO_2	Ar—NO_2
M − 48	SO	Aromatic sulphoxide
. . . .	. . . .	. . . .
M − 55	C_4H_7	Butyl ester
M − 56	C_4H_8	⎧ Ar—n-C_5H_{11}, ArO—n-C_4H_9 ⎨ Ar—iso-C_5H_{11}, ArO—iso-C_4H_9 ⎩ Pentyl ketone
M − 57	C_4H_9	Butyl ketone
M − 57	C_2H_5CO	Ethyl ketone
M − 58	C_4H_{10}	—
M − 60	CH_3COOH	Acetate

CI spectra

The power of CI in determining molecular weights of volatile compounds is illustrated in Fig. 4.25. Although the EI mass spectrum of dioctyl phthalate contains no peak due to M^{+}, abundant MH^{+} peaks are observed in both the CH_4 and isobutane CI spectra. In accord with the considerations in Sec. 4.2, the isobutane spectrum contains less abundant fragment ions. The formation of the fragment ions is rationalized in Scheme 4.4.

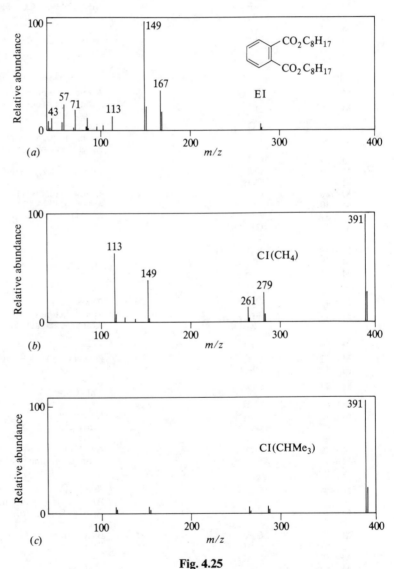

Fig. 4.25

(Reprinted with permission from Howe, Williams, and Bowen, 1981.)

Scheme 4.4

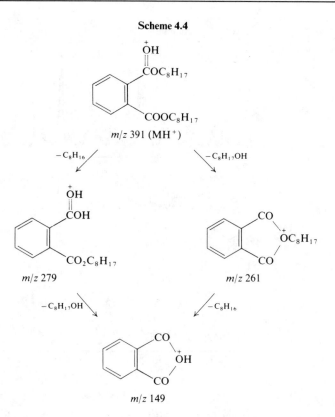

m/z 391 (MH$^+$)

m/z 279

m/z 261

m/z 149

Since CI spectra normally contain relatively abundant MH$^+$ ions, and relatively few fragment ions, CI is a technique well-suited to MS/MS work (Sec. 4.3). Some impressive analyses can be achieved using this method. In Fig. 4.26, data which indicate the presence of phenobarbital in a pharmaceutical preparation (Chardonna) are presented. The preparation was placed on the probe without any pretreatment. The collisional activation spectrum (Fig. 4.26a) of m/z 223 (the protonated molecular ion formed by methane chemical ionization) is very similar to that formed from authentic phenobarbital (Fig. 4.26b).

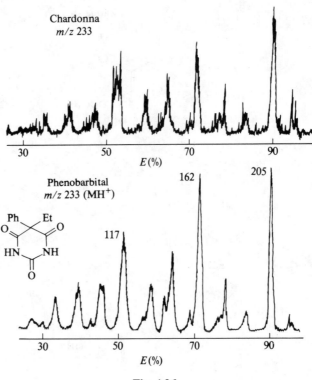

Fig. 4.26

(Reprinted with permission from Howe, Williams, and Bowen, 1981.)

FD spectra

FD gives spectra which are normally dominated by M^+ and/or MH^+. It is very useful for the molecular weight determination of relatively small, highly polar molecules (e.g. D-glucose, Fig. 4.27), or of molecules in the molecular weight range 500–3000 Daltons of intermediate polarity (i.e. polar and yet not sufficiently so to be readily soluble in glycerol for study by FAB or SIMS).

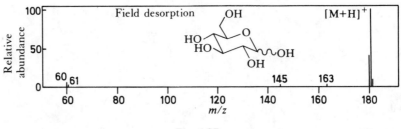

Fig. 4.27

(Reprinted with permission from Howe, Williams, and Bowen, 1981.)

FAB spectra

Salts (A$^+$B$^-$), and highly polar molecules of molecular weights up to 10 000 Daltons, are analysed well by this technique. The former point is illustrated by the positive-ion spectrum (Fig. 4.28) of a phosphonium bromide where A$^+$ (m/z 349) is the most abundant ion. In contrast, hydrocarbons and other highly hydrophobic molecules (e.g. steroids having little functionality) are not handled well by particle bombardment methods.

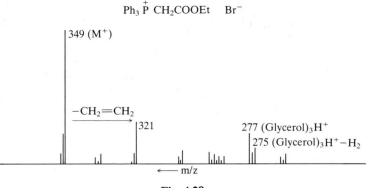

Fig. 4.28

The ability to determine easily the molecular weights of polar molecules has given a new aid to structure elucidation. A reaction is sought which is characteristic of a given functional group, and which can be carried out in high yield on a small (e.g. microgram) scale, preferably without tube-to-tube transfers. The difference in molecular weight before and after reaction indicates the number of such functionalities. Table 4.13 lists some useful examples.

Table 4.13 Selective microscale reactions of some common functional groups

Functional group	Reagent	Product	Change in mass per functional group
RNH_2	Ac_2O/H_2O (30 min)†	$RNHCOCH_3$	+42
ROH	Ac_2O/pyridine (overnight)	$ROCOCH_3$	+42
RCOOH	0.5% HCl in MeOH (overnight)	RCOOMe	+14
$RCONH_2$	$C_6H_5I(OCOCF_3)_2$	RNH_2	−28

† Reaction mixture buffered to pH ~ 8.5 with NH_4HCO_3.

Although multiply charged ions are occasionally seen in FAB mass spectra (e.g. MH_2^{2+} in spectra of peptides containing two arginine residues), the unfavourable nature of multiple ionic charges in the gas phase means that such ions are normally of low abundance even if present at all.

FAB mass spectrometry, in general, results in relatively little fragmentation. However, sufficient fragmentation is often seen in the mass spectra of peptides having molecular weights in the range 300–2000 Daltons to give sequence information.

Common fragmentations which may occur at a given amino acid residue are summarized in Scheme 4.5. In positive-ion spectra the fragments are recorded carrying an extra proton (H⁺), and in negative-ion spectra lacking a proton (with the exception of **17**, which is already charged and appears in positive ion spectra only).

Scheme 4.5

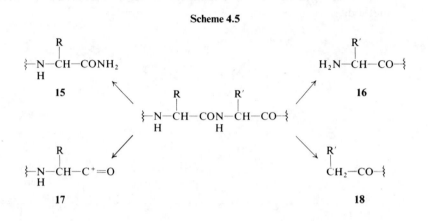

Ions of types **15–18** are found where the amino acid content of the ion differs by one residue. Therefore, given the atomic masses of the amino acids (Table 4.14), peptide sequences can be determined. This is illustrated by the negative-ion spectrum of a peptide toxin in Fig. 4.29, which gives the partial sequence X-Ile-Asp-Asp-Glu-Gln, and in conjunction with the positive-ion spectrum and the molecular weight, the total sequence:

C₆H₅CO-Ala-Phe-Val-Ile-Asp-Asp-Glu-Gln

$$C_6H_5CO\text{-Ala-Phe-Val-Ile-Asp-Asp-Glu-Gln}$$

Note that Leu and Ile (Table 4.14) have to be differentiated by amino acid analysis, and that although Lys and Gln have the same mass, they can be differentiated by acetylation of the former. Since FAB mass spectra contain a small peak at essentially every mass, reliable mass calibration of the spectrum is not usually a problem.

Table 4.14   Integral masses of amino acid residues $(-\overset{\displaystyle R}{\underset{\displaystyle H}{\mathrm{N-CH-CO}}}-)$†

Gly	57	Ile	113	Asn	114	Tyr	163	Pro	97
Ala	71	Ser	87	Glu	129	Trp	186	Met	131
Val	99	Thr	101	Gln	128	Lys	128	Arg	156
Leu	113	Asp	115	Phe	147	His	137	Cys	103

† When the first amino acid is lost from the carboxyl terminus of a peptide to give an ion of type **15**, the mass lost from (M—H)⁻ or MH⁺ is one mass unit greater than the values given in the table.

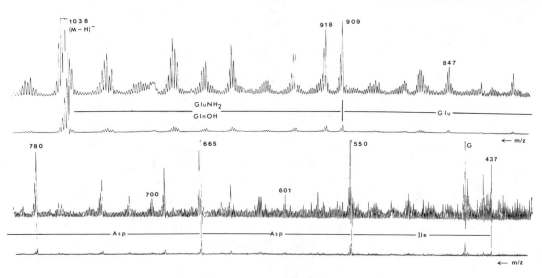

Fig. 4.29

4.5   Gas chromatography–mass spectrometry (GC/MS)

The separation and detection of components from a mixture of organic compounds is readily achievable by gas chromatography. Furthermore, limited characterization of unknown components is often possible from retention times appropriate to the particular column used. Mass spectrometry, because of its high sensitivity and fast scan speeds, is the technique most suited to provide definite structural information from the small quantities of material eluted from a gas chromatograph. The association of the two techniques has, therefore, provided a powerful means of structure identification for the components of natural and synthetic organic mixtures. Mass spectra of acceptable quality are potentially obtainable for every component that may be separated by the gas chromatograph, even though the components may be present in nanogram quantities and eluted over periods of only a few seconds.

A schematic diagram of a GC/MS system is shown in Fig. 4.30.

The interface between the GC and the MS in Fig. 4.30 is a jet separator. It is necessary to introduce this component if relatively large quantities of carrier gas are leaving the GC; if all the carrier gas were allowed to enter an EI source, the pressure rise in the source would be unacceptable. Using a jet separator, most of the carrier gas is pumped away (by utilizing the relatively fast diffusion of the low molecular weight carrier gas) into a pumped interface between two aligned orifices (typically of diameter 50–100 μm). The necessity for a separator can be avoided if capillary GC (low volumes of carrier gas) is employed in conjunction with CI (high source pressures).

In GC/MS, the mass spectrometer can also be used as a detector. The total ion current produced in the mass spectrometer is used instead of flame ionization or electron capture detection. In view of the large amounts of data that are generated by

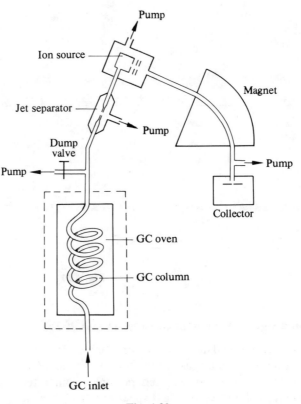

Fig. 4.30

(Reprinted with permission from Howe, Williams, and Bowen, 1981.)

a GC/MS, it is necessary that the system be coupled on-line to a computer-controlled data system (Sec. 4.7).

4.6   Liquid chromatography–mass spectrometry (LC/MS)

High-pressure liquid chromatography (HPLC) is a powerful method for the separation of complex mixtures, especially when many of the components may have similar polarities. In reverse-phase HPLC, the column substrate is such that starting with an aqueous solution of a mixture of polar components, the most polar components are eluted first. The later-eluted hydrophobic components are often encouraged to leave the column by gradually increasing the concentration of acetonitrile (CH_3CN) in the otherwise aqueous developing solvent.

If a mass spectrum of each component can be recorded as it elutes from the LC column, quick characterization of the components is greatly facilitated. The problems to be overcome are those of removing relatively large amounts of polar solvents, and of maintaining polar solutes in the gas phase as ions. To date, the most successful method of doing this is the thermospray technique in which the LC effluent progresses into an 0.15 mm internal diameter stainless steel capillary which is surrounded by a heated copper block (Fig. 4.31). When the block is heated to temperatures in the range 100–350°C (depending on the solvent), the eluate can be induced to vaporize just as it reaches the exit of the capillary tube. Under these circumstances the vaporizer produces a superheated mist carried in a supersonic jet of vapour. If the eluate contains solutes which have functionalities that readily maintain a charge (e.g. —NH_2 or —COOH), and especially if it contains a volatile buffer which can promote the formation of ions (e.g. 0.1 M NH_4OAc or HCO_2H), the droplets in the mist may be positively or negatively charged. Therefore, when the solvent evaporates from these droplets, MH^+ or $(M—H)^-$ may remain in the gas phase to be recorded.

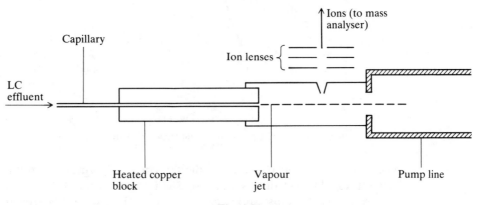

Fig. 4.31

As shown in Fig. 4.31, solvent molecules are removed by fast pumping, whereas ions are extracted from the jet by suitable electric fields and then injected into the mass analyser. The system can handle flow rates of an aqueous solvent up to 2 ml min^{-1}, and good mass spectra of amino acids, peptides, nucleotides, and antibiotics up to molecular weights of nearly 2000 Daltons have been obtained. Note that since the ions may already be present in solution, often no ionizing device is necessary in LC/MS coupling by the thermospray technique.

4.7 MS data systems

If a computerized data system is incorporated on-line to a mass spectrometer, the system can be used not only for the storage, manipulation, and retrieval of data, but also to control (through the computer keyboard) the GC and the mass spectrometer. Examples of the latter control include automatic sampling, injection, and scanning; and control of the gas flow rate in the GC, and the temperature gradient (if any) to be applied to the GC column.

The normal output of a mass spectrometer is a voltage which varies continuously as the spectrum is scanned. The continuous electrical output (i.e. analogue signal) from the detector to the mass spectrometer is sampled at precise intervals (Fig. 4.32) and converted into pulsed (i.e. digital) form by an analogue-to-digital (A/D) converter. The resulting discrete values are suitable for handling by the computer.

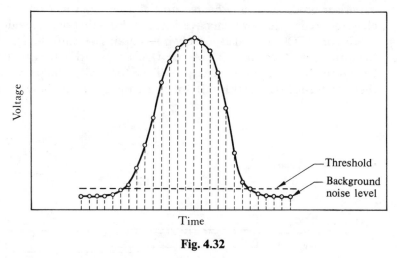

Fig. 4.32

(Reprinted with permission from Howe, Williams, and Bowen, 1981.)

The next step is the rejection of digital samples having an amplitude below a selected threshold level, or less than a selected time width. In this way, random noise and narrow electrical noise spikes can be rejected (Fig. 4.32). The time of arrival of a peak centroid is converted to mass by calibrating the system using the arrival time of peaks of known mass. All the processes which have to be gone through before storage of the digitized spectrum on disc are summarized in Fig. 4.33. The interscan report allows the user to monitor spectra by displaying the data on a visual display unit (VDU) as a GC scan progresses. An obvious advantage of this procedure is that the experiment can be modifed as it proceeds if the spectra obtained are unsatisfactory.

The computer may also be used to manipulate the data. For example, 'background subtraction' (e.g. the subtraction of peaks due to column bleed, or other impurities) results in the presentation of cleaner spectra. Such spectra may then be automatically compared (in terms of the mass and abundance of each peak) with a library of many thousands of mass spectra already stored on disc in the computer. This process of 'library searching' is an extremely powerful method for the identification of compounds whose mass spectra have previously been recorded. The identification is achieved by using the mass spectrum as a fingerprint.

4.8   Specific ion monitoring and quantitative MS

Although the electron multiplier which is normally used to detect ions after mass analysis is an extremely sensitive device (a useful signal can be obtained from as few as 20 ions), the scanning of magnetic sector and quadrupole mass spectrometers is

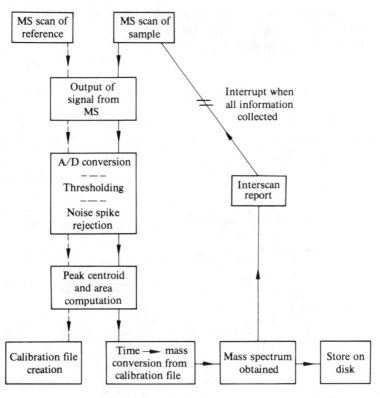

Fig. 4.33

(Reprinted with permission from Howe, Williams, and Bowen, 1981.)

inefficient in terms of signal-to-noise ratio (S/N). Ions are typically recorded for $\leqslant 1$ per cent of the time they are produced. If sensitivity is of supreme importance, then S/N can be improved by monitoring only the most abundant ion or ions. The former process is known as single-ion monitoring (SIM) and the latter as multiple-ion monitoring (MIM). In a magnetic sector instrument, the desired result is achieved by setting the magnet current to scan repetitively through a single-ion signal (SIM); or for the current to alternate between values which lead to the recording of a few abundant (and/or) characteristic ions (MIM). In the latter case, the signal from each ion is recorded by a different channel of a multi-channel recorder.

An application of MIM to the identification in human blood of metabolites of the drug chlorpromazine (**19**) is illustrated in Fig. 4.34. The mass spectra of chlorpromazine and its sidechain derivatives contain abundant ions at m/z 246 (**20**), 232 and 234 (**21**). These three ions were monitored continuously from the GC effluent of the derivatized blood extract (Fig. 4.34). The abundance of the three ions rose and fell simultaneously for two of the eluted compounds, strongly suggesting, therefore, that these eluants were chlorpromazine sidechain derivatives. The structures were confirmed by monitoring suspected molecular ions and utilizing authentic GC retention times. The two derivatives were, in fact, the trifluoracetates of des- and di-desmethylchlorpromazine.

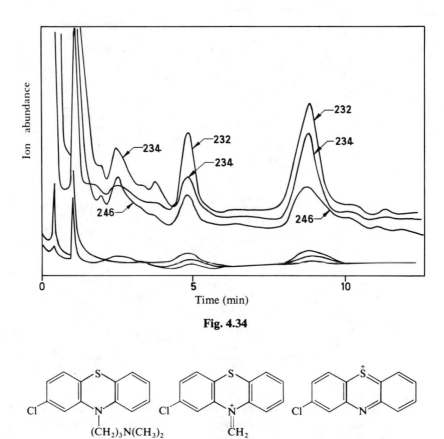

Fig. 4.34

19

20, *m/z* 246

21, *m/z* 232

The MIM technique is both remarkably specific and sensitive; in favourable cases, compounds can be identified at the picogram level. The GC/MS/data system combination is extremely powerful in the analysis of drug metabolites, flavours, and perfumes.

On occasions, it is desirable to be able to quantitate a compound by mass spectrometry, at say the nanogram–picogram level. Such work requires a suitable internal standard, and an isotopically labelled analogue is usually used. For example, a mass spectrometer can readily separate ions produced from the pesticide methyl parathion (**22**) and its deuterated analogue (**23**). Thus, if an unknown quantity of **22** is spiked with a known quantity of **23**, the relative signal sizes permit quantitation of **22**.

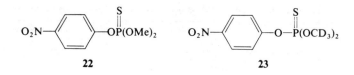

4.9   Interpreting the spectrum of an unknown

When working on real laboratory problems, it is important to know whether the sample is of high purity, or impure. Useful information can come from the sharpness, and constancy, of the melting point (in the case of crystalline, or recrystallized samples), TLC, GC, and LC. Samples which have been extensively handled (e.g. on thin-layer plates, columns, and greased apparatus) may contain characteristic impurity peaks (Table 4.11). Since molecular ions of aliphatic compounds are frequently of low abundance in EI spectra, it is important in these cases to determine, in addition, a spectrum using a 'soft' ionization technique (CI, FD, FAB, SIMS, plasma desorption; Sec. 4.2). For the identification in a mixture of a volatile 'unknown' whose structure has been determined in earlier work, GC/MS (Sec. 4.5) with subsequent searching of a computer file of EI spectra (Sec. 4.7) is a powerful method.

At an early point in the structure elucidation, the molecular ion must be identified with certainty. Ensure that the supposed molecular ion (in EI and/or 'soft' ionization spectra) is separated from peaks at lower masses by acceptable mass differences (Table 4.12, and Sec. 4.4). Note whether the m/z value of the molecular ion is odd or even (Sec. 4.4) and any characteristic isotope patterns (Sec. 4.4).

Once a molecular ion [M^+, MH^+, or $(M—H)^-$] has been identified with certainty, decide whether it is necessary to determine (or limit) the molecular formula of the compound by a high-resolution measurement of the mass of this molecular ion (Sec. 4.3). Note that such a measurement (of a given precision in p.p.m.) will be consistent with increasing numbers of combinations of elements as the molecular weight increases.

After the molecular weight (and possibly molecular formula also) has been determined, use any available UV, IR, and NMR spectra, and chemical knowledge to determine a suggested partial or total structure. At this stage, return to the mass spectrum to see if the observed fragmentation pattern is consistent with the proposal. In checking this consistency, remember that observed single-bond cleavages must correspond to the energetically most favourable combinations of products (Sec. 4.4).

In view of the enormous variety of ion masses which can arise from an equally enormous variety of organic compounds, the recognition of structural units from m/z values is of limited value even for molecules of molecular weights $\leqslant 200$. Nevertheless, for such relatively small molecules where the nature of the functionality is limited and/or known, it is probably worthwhile to check the observed m/z values with those given in Table 4.15.

Table 4.15 Masses and some possible compositions of common fragment ions

m/z	Groups commonly associated with the mass	Possible inference
15	CH_3^+	—
18	H_2O^+	—
26	$C_2H_2^+$	—
27	$C_2H_3^+$	—
28	$CO^+, C_2H_4^+, N_2^+$	—
29	$CHO^+, C_2H_5^+$	—
30	$CH_2{=}\overset{+}{N}H_2$	Primary amine?
31	$CH_2{=}\overset{+}{O}H$	Primary alcohol?
36/38(3:1)	HCl^+	—
39	$C_3H_3^+$	—
40†	$Argon^+, C_3H_4^+$	—
41	$C_3H_5^+$	—
42	$C_2H_2O^+, C_3H_6^+$	—
43	CH_3CO^+	CH_3COX
43	$C_3H_7^+$	C_3H_7X
44	$C_2H_6N^+$	Some aliphatic amines
44	$O{=}C{=}\overset{+}{N}H_2$	Primary amides
44	$CO_2^+, C_3H_8^+$	—
44	$CH_2{=}CH(OH)^+$	Some aldehydes
45	$CH_2{=}\overset{+}{O}CH_3$ $CH_3CH{=}\overset{+}{O}H$	} Some ethers and alcohols
47	$CH_2{=}\overset{+}{S}H$	Aliphatic thiol
49/51(3:1)	CH_2Cl^+	—
50	$C_4H_2^+$	Aromatic compound
51	$C_4H_3^+$	C_6H_5X
55	$C_4H_7^+$	—
56	$C_4H_8^+$	—
57	$C_4H_9^+$	C_4H_9X
57	$C_2H_5CO^+$	{ Ethyl ketone Propionate ester
58	$CH_2{=}C(OH)CH_3^+$	{ Some methyl ketones Some dialkyl ketones
58	$C_3H_8N^+$	Some aliphatic amines
59	$COOCH_3^+$	Methyl ester
59	$CH_2{=}C(OH)NH_2^+$	Some primary amides
59	$C_2H_5CH{=}\overset{+}{O}H$	$C_2H_5CH(OH){-}X$
59	$CH_2{=}\overset{+}{O}{-}C_2H_5$ and isomers	Some ethers
60	$CH_2{=}C(OH)OH^+$	Some carboxylic acids
61	$CH_3CO(OH_2)^+$	$CH_3COOC_nH_{2n+1}(n>1)$
61	$CH_2CH_2SH^+$	Aliphatic thiol
66	$H_2S_2^+$	Dialkyl disulphide
69	CF_3^+	—
68	$CH_2CH_2CH_2CN^+$	—
69	$C_5H_9^+$	—
70	$C_5H_{10}^+$	—
71	$C_5H_{11}^+$	$C_5H_{11}X$
71	$C_3H_7CO^+$	{ Propyl ketone Butyrate ester

† Appears as a doublet in the presence of argon from air; useful as a reference point in counting the mass spectrum.

Table 4.15 *continued*

m/z	Groups commonly associated with the mass	Possible inference
72	$CH_2=C(OH)C_2H_5{}^+$	Some ethyl alkyl ketones
72	$C_3H_7CH=\overset{+}{N}H_2$ and isomers	Some amines
73	$C_4H_9O^+$	—
73	$COOC_2H_5{}^+$	Ethyl ester
73	$(CH_3)_3Si^+$	$(CH_3)_3SiX$
74	$CH_2=C(OH)OCH_3{}^+$	Some methyl esters
75	$(CH_3)_2Si=\overset{+}{O}H$	$(CH_3)_3SiOX$
75	$C_2H_5CO(OH_2)^+$	$C_2H_5COOC_nH_{2n+1}\ (n>1)$

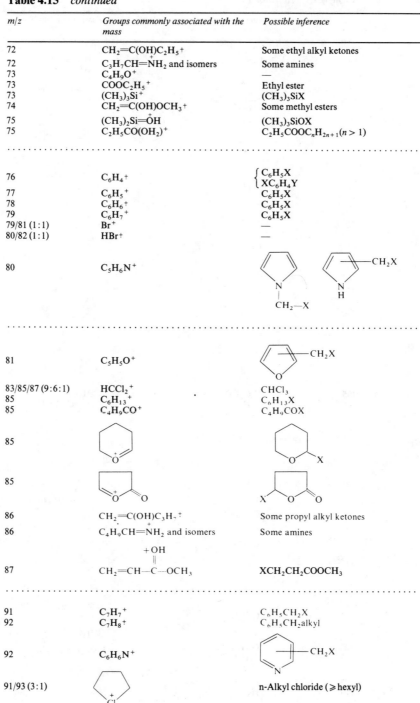

76	$C_6H_4{}^+$	$\begin{cases} C_6H_5X \\ XC_6H_4Y \end{cases}$
77	$C_6H_5{}^+$	C_6H_5X
78	$C_6H_6{}^+$	C_6H_5X
79	$C_6H_7{}^+$	C_6H_5X
79/81 (1:1)	Br^+	—
80/82 (1:1)	HBr^+	—
80	$C_5H_6N^+$	

81	$C_5H_5O^+$	
83/85/87 (9:6:1)	$HCCl_2{}^+$	$CHCl_3$
85	$C_6H_{13}{}^+$	$C_6H_{13}X$
85	$C_4H_9CO^+$	C_4H_9COX
85		
85		
86	$CH_2=C(OH)C_3H_7{}^+$	Some propyl alkyl ketones
86	$C_4H_9CH=\overset{+}{N}H_2$ and isomers	Some amines
87	$CH_2=CH-\overset{\overset{+OH}{\|}}{C}-OCH_3$	$XCH_2CH_2COOCH_3$

91	$C_7H_7{}^+$	$C_6H_5CH_2X$
92	$C_7H_8{}^+$	$C_6H_5CH_2alkyl$
92	$C_6H_6N^+$	
91/93 (3:1)		n-Alkyl chloride ($\geqslant$ hexyl)

Table 4.15   *continued*

m/z	Groups commonly associated with the mass	Possible inference
93/95 (1:1)	CH_2Br^+	—
94	$C_6H_6O^+$	C_6H_5O-alkyl (alkyl $\neq CH_3$)
94		
95		
95	$C_6H_7O^+$	
97	$C_5H_5S^+$	
99		
99		
105	$C_6H_5CO^+$	C_6H_5COX
105	$C_8H_9{}^+$	CH_3—$C_6H_4CH_2X$
106	$C_7H_8N^+$	
107	$C_7H_7O^+$	
107/109 (1:1)	$C_2H_4Br^+$	
111		
121	$C_8H_9O^+$	
122	C_6H_5COOH	} Alkyl benzoates
123	$C_6H_5COOH_2{}^+$	
127	I^+	—
128	HI^+	—

Table 4.15 *continued*

m/z	Groups commonly associated with the mass	Possible inference
135/137 (1:1)		n-Alkyl bromide (⩾ hexyl)
130	$C_9H_8N^+$	
141	CH_2I^+	—
147	$(CH_3)_2Si\!=\!\overset{+}{O}\!-\!Si(CH_3)_3$	—
149		Dialkyl phthalate
160	$C_{10}H_{10}NO^+$	
190	$C_{11}H_{12}NO_2{}^+$	

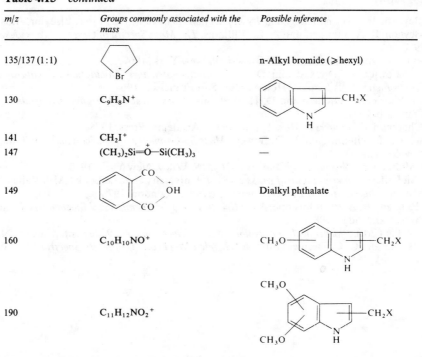

Bibliography

J. H. Beynon, *Mass Spectrometry and Its Applications to Organic Chemistry*, Elsevier, Amsterdam, 1960.

J. H. Beynon, R. A. Saunders and A. E. Williams, *The Mass Spectra of Organic Molecules*, Elsevier, London, 1968.

K. Biemann, *Mass Spectrometry*, McGraw-Hill, New York, 1962.

H. Budzikiewicz, C. Djerassi and D. H. Williams, *Structure Elucidation of Natural Products by Mass Spectrometry*, Vols I and II, Holden-Day, San Francisco, 1964.

H. Budzikiewicz, C. Djerassi and D. H. Williams, *Mass Spectra of Organic Compounds*, Holden-Day, San Francisco, 1967.

J. R. Chapman, *Computers in Mass Spectrometry*, Academic Press, 1978.

I. Howe, D. H. Williams, and R. D. Bowen, *Mass Spectrometry—Principles and Applications*, McGraw-Hill, New York, 1981.

W. H. McFadden, *Techniques of Combined GC/MS*, Wiley, New York, 1973.

F. W. McLafferty, *Interpretation of Mass Spectra*, University Science Books, Mill Valley, California, 1980.

B. J. Millard, *Quantitative Mass Spectrometry*, Heyden, London, 1978.

M. E. Rose and R. A. W. Johnstone, *Mass Spectrometry for Chemists and Biochemists*, Cambridge University Press, Cambridge, 1982.

G. R. Waller (Ed.), *Biochemical Applications of Mass Spectrometry*, Wiley-Interscience, New York, 1972.

G. R. Waller and O. C. Dermer (Eds), *Biochemical Applications of Mass Spectrometry*, Wiley-Interscience, New York, 1980.

5. Structure elucidation by joint application of UV, IR, NMR, and mass spectroscopy

5.1 *General approach.* 5.2 *Worked examples.* 5.3 *Problems.*

5.1 General approach

Between them, the four spectroscopic methods give an immense amount of information about the structure of an organic molecule, often leading to an unambiguous structure. The best order in which to take each piece of information is not always quite the same; nor is it always obvious. In this chapter, we shall work through four examples to show how it can be done, and at the end of the chapter there are 18 problems for you to work through. Each of them gives an unambiguous answer, but be prepared to find that in real life this is not always going to be the case. For a longer discussion of how to go about a structure determination, see Kemp, *Qualitative Organic Analysis*, McGraw-Hill, London, 2nd Ed., 1986.

Frequently the best place to begin is with the molecular ion in the mass spectrum, from which a molecular formula may be deduced. In the absence of a molecular ion, a combustion analysis also gives this information. It is then wise quickly to work out the number of double bonds and rings in the molecule, in order to get an idea of the degree of complexity in the structure. This is done by inspecting the molecular formula. If the molecule contains only C, H, and O, then the number of double bonds and rings (double bond equivalents, DBE) is given by Eq. 5.1.

$$C_aH_bO_c \qquad DBE = \frac{(2a + 2) - b}{2} \tag{5.1}$$

The $(2a + 2)$ term is the number of hydrogens in a saturated hydrocarbon having a carbon atoms. Since every ring or double bond means two fewer hydrogen atoms (cyclohexane is C_6H_{12} and ethylene, C_2H_4), subtracting b, the actual number of hydrogen atoms present, from $(2a + 2)$ and dividing by two gives the total number of double bonds and rings in the molecule. It is useful to remember that a benzene ring has a total of four double bond equivalents: three 'double bonds' and one ring. The number of divalent atoms (O, S, etc.) present makes no difference to this sum, but mono- and trivalent atoms do. Count monovalent atoms (Cl, Br, etc.) as hydrogens and add them to b. When trivalent atoms (N, trivalent P, etc.) are present, use Eq. 5.2, i.e. subtract one from b for each trivalent atom present.

$$C_aH_bO_cN_d \qquad DBE = \frac{(2a + 2) - (b - d)}{2} \tag{5.2}$$

Thus the formula $C_5H_{11}N$ has one double bond equivalent (often written: $\overline{1}$); it might be the i-propylimine of acetaldehyde (one double bond) or cyclopentylamine (one ring). Many other structures are possible, but we can already see that, if one of the spectroscopic methods reveals the presence of a double bond (C=C, C=O, or C=N), then there are no rings. Conversely, if double bonds are clearly not present, there must be a ring. It is very helpful to carry this information with you into the next stage of the problem, which is usually to look very briefly at the UV spectrum. If you have deduced that there are two or more double bond equivalents and the UV shows strong absorption, then two double bonds conjugated to each other are present. At this stage, it is not usually helpful to seek more information than this from the UV spectrum; it is better to turn to the IR and identify the functional groups present. The NMR spectra (^{13}C and ^{1}H) will then enable you to arrange the carbon and hydrogen skeleton and to join on the functional groups as far as possible. The mass spectrum may also help by identifying fragments of the molecule, but it is at this stage that there is no fixed order in which to put together all the information.

With experience and a familiarity with all the data in Chapters 1–4, you will find it is often easy (and enjoyable) to put together a structure for an unknown compound using only the four spectroscopic methods. The best way to practise with the four examples that follow is to see how much of the structure you can deduce for yourself from the spectra, and only then read our analysis. In all four examples and the first 12 problems that follow, the ^{1}H NMR spectra are taken at 100 MHz, so that the small scale divisions are 10 Hz (this may help you when you want to estimate the magnitude of coupling constants), and the ^{13}C NMR spectra are all taken in CDCl$_3$, which, in several cases, gives rise to three peaks of nearly equal intensity between δ 75 and 79. In Problems 13–18, the ^{1}H NMR spectra are taken at 250 MHz. All mass spectra are EI mass spectra.

5.2 Worked examples

Example 1

The mass spectrum and the combustion analysis show that the formula is C_4H_8O, and there is therefore one double bond equivalent. The UV spectrum has λ_{max} 295 nm, but before doing anything with this information, it is important always to work out the ε value from Eq. 1.2 in order to find out how intense the absorption is. Equation 1.2 is usefully rewritten as Eq. 5.3.

$$\varepsilon = \frac{absorbance \times molecular\ weight \times 100}{weight\ of\ compound\ in\ 100\ ml \times path\ length\ in\ cm} \tag{5.3}$$

In this case the numbers are

$$\varepsilon = \frac{0.28 \times 72 \times 100}{106 \times 1} = 19$$

The absorption is therefore very weak, and typical of the $n \rightarrow \pi^*$ absorption of a saturated ketone or aldehyde (Sec. 1.17). One should not place too much faith in this information alone—the absorption could so easily be a trace of strongly absorbing impurity, which the other spectroscopic methods would not detect. However, in this case, the presence of a carbonyl group is immediately apparent in the IR spectrum with its very strong band at 1715 cm^{-1}; furthermore, it is clearly a ketone and not an aldehyde, since the latter would have: a carbonyl band at slightly higher frequency (Table 2.10); further absorption between 2900 and 2700 cm^{-1} (Table 2.2); and absorption in the δ 9–10 region of the ^{1}H NMR spectrum. With only one double bond equivalent and one heteroatom to account for, we have now found them both in the ketone carbonyl group. There can be no further functionality, so all we have to do is arrange the carbon skeleton. Here we could look equally profitably at either the NMR or MS. The ^{13}C NMR shows that all four carbon atoms are different; one (at δ 208.79) is clearly the carbonyl carbon, being both weak and at a very low field (Table 3.13). As it happens, in this very simple case, we already have all the information we need with which to deduce a structure, but typically we would look next at the ^{1}H NMR, where we see a quartet (δ 2.4), a singlet (2.09) and a triplet (1.04). The integration trace gives a rise of 1.6 cm for the quartet, 2.1 cm for the singlet, and 2.2 cm for the triplet. Since there is a total of eight hydrogen atoms in the molecule, these correspond to 2.2H, 2.8H, and 3.0H, respectively; clearly the true ratios are 2:3:3. The three-proton signal at δ 2.09 will be a methyl group; since it is a singlet, it must be attached to an atom having no protons on it. Its position of resonance is compatible (Table 3.17) with its being a $CH_3CO—$ group. The 1:3:3:1 quartet and the 1:2:1 triplet integrating for two and three protons, respectively, is typical of an ethyl group. Since the former is only a quartet, the CH_2 group must be attached to an atom having no hydrogen atoms on it, and since it is in the δ 1.4–2.4 range (Table 3.17, column 2), that atom must be carbon. Clearly it is joined directly to the carbonyl carbon, and the whole structure is, of course, methyl ethyl ketone, $CH_3COCH_2CH_3$.

Although we have fully solved the structure at this stage, it is always wise to look at the other spectra and confirm that we have not been misled. The positions of the ^{13}C signals, for instance, are obviously right for this structure: the peak at 7.86 is the isolated methyl group, $CH_3COCH_2CH_3$, the peak at 29.37 is the methyl group next to the ketone, $CH_3COCH_2CH_3$, and the peak at 36.80 is the methylene group, $CH_3COCH_2CH_3$ (Table 3.6). These assignments are confirmed by the off-resonance spectrum (section 3.6), in which the same signals are a quartet, a quartet, and a triplet, showing that the corresponding carbon atoms carry three, three, and two hydrogens, respectively. Note that the carbonyl signal is still a singlet in the off-resonance spectrum.

Example 1   Found: C, 66.7%
                     H, 11.1%

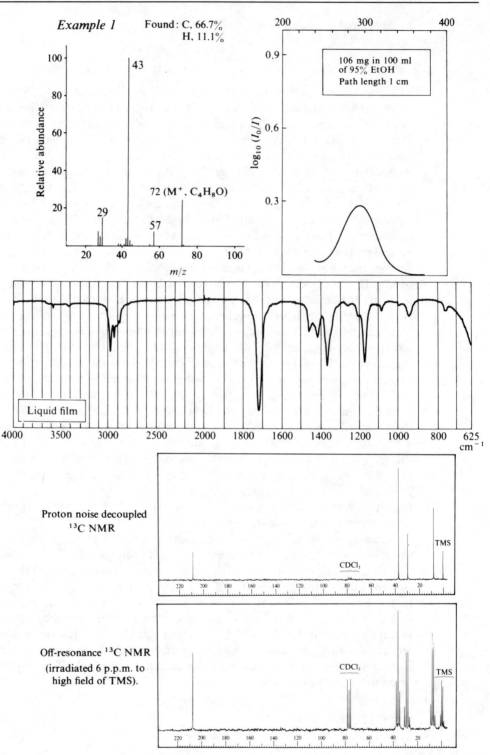

Proton noise decoupled
^{13}C NMR

Off-resonance ^{13}C NMR
(irradiated 6 p.p.m. to
high field of TMS).

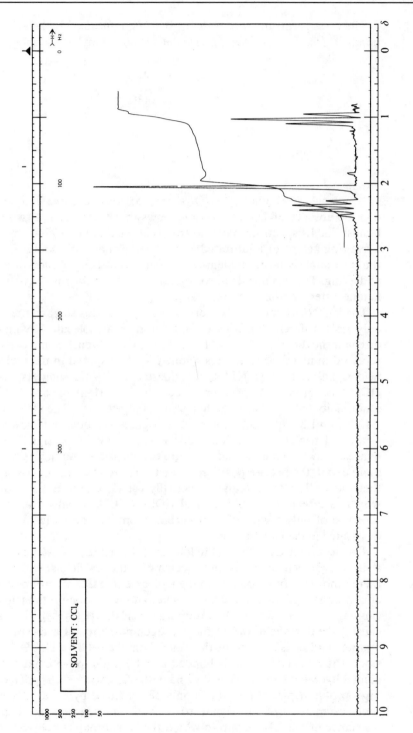

SOLVENT: CCl₄

The mass spectrum, too, is definitive: the molecule breaks at each of the C—CO bonds (**1**) (see Table 4.8) and hence detects the methyl and ethyl groups directly.

1

Example 2

The molecular formula $C_{11}H_{20}O_4$ shows that there are two double bond equivalents and the absence of UV absorption shows that these are not two *conjugated* double bonds. The IR spectrum shows a strong carbonyl band at 1740 cm^{-1}, which could be a five-ring ketone or a saturated ester. There is no C=C double bond absorption, so the two double bond equivalents are either two carbonyl groups or one carbonyl group and a ring. There is no OH absorption in the IR, so the four oxygen atoms must be in ketone, ester, or ether groups.

The ^{13}C NMR spectrum is very informative at this stage in the analysis—it shows only *eight* different kinds of carbon atom in a molecule having *eleven* in all (we discount the three small peaks from the CDCl$_3$ solvent). Some carbon atoms must be repeated in identical structures symmetrically disposed in the molecule. The lowest-field signal in the ^1H NMR, the quartet at δ 4.12, suggests the presence of an OCH_2CH$_3$ group—its position of resonance is right for an OCH$_2$ group and the multiplicity tells us that a methyl group is joined to it. The OCH$_2$CH_3 signal is also visible at δ 1.3, but it is on top of other signals and does not show up quite as clearly as a 1:2:1 triplet as it might. The intensity of the OCH_2 signal corresponds to four hydrogens, which means that there are two (possibly identical) OCH$_2$CH$_3$ groups. In more detail, the precise position of the OCH_2 resonance (4.12) suggests (Table 3.17) that the OCH$_2$CH$_3$ groups are actually esters, COOCH$_2$CH$_3$, and not ethers, in which case there are two identical COOCH$_2$CH$_3$ groups, thus accounting for the presence of only eight different carbon atoms in the molecule (three of them are duplicated in the CO$_2$Et groups).

The next informative signal to look at is the triplet at δ 3.2: integration makes this a single hydrogen and we have to account for the low field at which it resonates. Since there is no other functionality in the molecule (the CO$_2$Et groups account for the two double bond equivalents and all the heteroatoms), a one-proton triplet at this position can only be produced by the grouping —CH$_2$CH(CO$_2$Et)$_2$, the carbethoxy groups causing the downfield shift of the hydrogen near them. Furthermore, the two-proton quartet at δ 1.88 is likely to be the signal from the —CH_2CH(CO$_2$Et)$_2$ hydrogen, and, since it is a quartet, it too is bonded to a CH$_2$ group (to make the total number of vicinal hydrogens three). Thus we have the fragment —CH$_2$CH$_2$CH(CO$_2$Et)$_2$. The three-proton triplet at δ 1.00 can only be produced by a CH_3CH$_2$— group, and the slight asymmetry of the signal arises from the nearness in chemical shift of the resonance of the CH$_2$ group to which the CH$_3$ group is coupled; the CH$_2$ signal is

Example 2   Found: C, 61.0%
             H, 9.4%

No UV maximum
above 200 nm

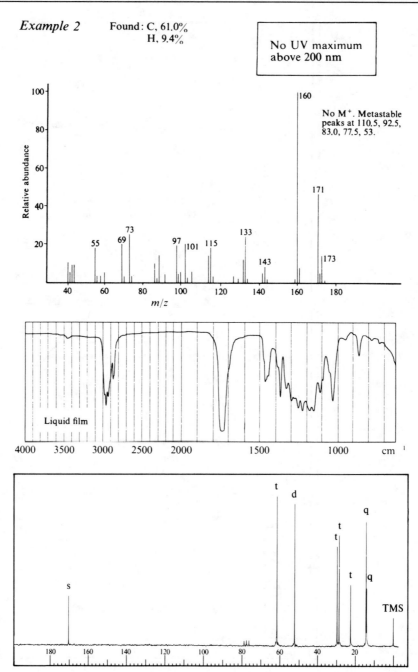

No M$^+$. Metastable
peaks at 110.5, 92.5,
83.0, 77.5, 53.

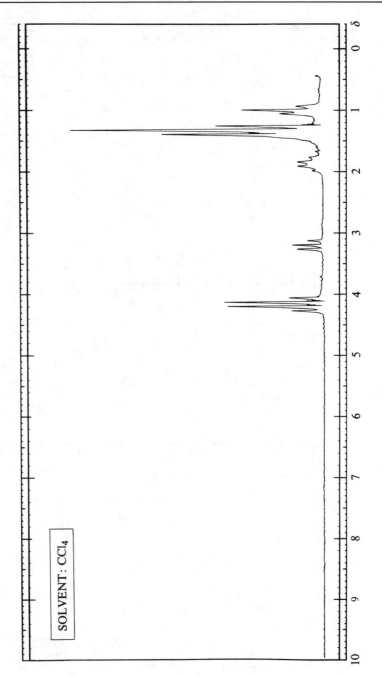

SOLVENT: CCl$_4$

probably under the triplet at δ 1.3. The asymmetric triplet at δ 1.0 is typical of a CH_3CH_2 group at the end of an alkyl chain. The two fragments we have now identified, CH_3CH_2— and —$CH_2CH_2CH(CO_2Et)_2$, account for all the atoms of the molecular formula, and the structure must be made by joining them together as in **2**.

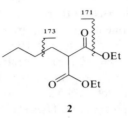

2

This structure also fits the ^{13}C NMR.

1. There are two different kinds of methyl group (at δ 13.81 and 14.10), with the signal at a lower field about twice as intense (because it comes from two identical methyl groups). We know that they are methyl groups, not only because of their chemical shift but also because they are quartets in the off-resonance spectrum. To save space, the off-resonance spectrum is not shown in this or any of the spectra from now on, but the multiplicity is shown on the completely decoupled spectra as a letter (s = singlet, d = doublet, t = triplet and q = quartet).
2. There are three C—CH_2—C groups (triplets in the off-resonance spectrum) at 22.38, 28.49, and 29.53. We discussed the assignment of these signals on page 128.
3. There is a methine group (a doublet) at 52.03.
4. There is the CH_2O carbon at 61.12 and the carbonyl carbon at 169.32.

The mass spectrum also confirms the structure **2**, and the base peak (m/z 160) is the result of β-cleavage with γ-hydrogen rearrangement (**3**) (Sec. 4.4, page 175).

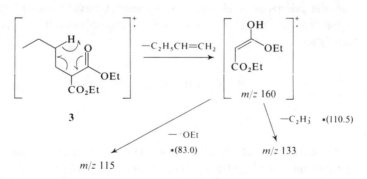

Example 3

The mass spectrum has no molecular ion, but combustion analysis gives a formula, $C_5H_{11}NO_4$. There is therefore one double bond equivalent, which is clearly not a carbonyl group because there is no absorption in the IR spectrum between 1900 and 1600 cm^{-1}. Instead, there is very strong absorption at 1545 cm^{-1}, which is likely to be from a nitro group (Table 2.16), a formulation supported first by the UV spectrum with a weak n → π* band at 275 nm (ε 24), and second by the absence of the molecular ion in the mass spectrum, which is common with aliphatic nitro compounds. The IR spectrum, with strong absorption at 3350 cm^{-1}, also shows the unmistakable presence of a hydroxyl group.

The ^{13}C NMR shows that there are only four different kinds of carbon atom; two carbons are in the same magnetic environment and are most likely to be two identical groups symmetrically disposed in the molecule. The next most useful piece of information is from the ^{1}H NMR spectrum. The integration distributes the eleven hydrogens thus: *four* hydrogens to the complicated signals centred at δ 4.12, *two* hydrogens to the triplet at 3.32, *two* hydrogens to the quartet at 2.00, and *three* hydrogens to the triplet at 1.00. These last two signals (the two-proton quartet and the three-proton triplet) are obviously an ethyl group, and the chemical shift and multiplicity of the quartet means it must be bound to a fully substituted carbon atom. In the off-resonance ^{13}C NMR, the *C*-ethyl group is responsible for the quartet at 7.66 (—CH$_2$CH$_3$) and either the triplet at 25.77 or the triplet at 63.51 (—CH$_2$CH$_3$). The fully substituted carbon atom must be the weak singlet at 94.23 (fully substituted carbons are usually weak, see Sec. 3.4). Thus the two identical carbon atoms must be responsible for whichever of the two triplets (25.77 and 63.51) is not produced by the CH$_2$ group of the ethyl group. One of these (63.51) is nearly twice as intense as the other, so it is likely that it comes from the two identical groups and the other (25.77) from the ethyl group. The two identical groups are therefore CH$_2$ groups (because they give rise to triplets in the off-resonance spectrum) and they are at a comparatively low field, suggesting that perhaps they are attached to a heteroatom (Table 3.6).

When the ^{1}H NMR spectrum is taken after a shake with D$_2$O (Sec. 3.5, page 75) (see the superimposed trace), the two-proton triplet at 3.32 disappears. This signal has therefore been produced by *two* OH groups, and the OH groups must be bonded to CH$_2$ groups to account for their appearance as a triplet. It is now clear that the two identical groups are CH$_2$OH groups, and we have found the fragments **4**, **5**, and **6**. These account for all the atoms, and there is only one way of putting them together, namely as **7**.

$$—NO_2 \qquad Y—\underset{Z}{\overset{X}{\underset{|}{\overset{|}{C}}}}—CH_2CH_3 \qquad (—CH_2OH)_2 \qquad HO\diagdown\overset{NO_2}{\diagup}OH$$

$$\textbf{4} \qquad\qquad \textbf{5} \qquad\qquad\qquad \textbf{6} \qquad\qquad\qquad \textbf{7}$$

A remarkable feature of the ^{1}H NMR spectrum still remains to be examined: the signal from the C*H*$_2$OH hydrogens. After the D$_2$O shake this is an AB quartet (δ_A 4.00, δ_B 4.24, J_{AB} = 12 Hz). The CH$_2$OH groups are bonded to a prochiral centre, *not* a chiral centre, but the A and B hydrogens of each of the —CH$_A$H$_B$OH groups 'see' three *different* groups on the adjacent carbon atom. They do not necessarily experience the same magnetic environment and can, as in this case, resonate with different chemical shifts. In that case, they couple with each other, giving rise to the AB system. Before the D$_2$O shake, they also couple with the OH proton, as can be

Example 3   Found: C, 35.2%
              H, 8.2%
              N, 10.1%

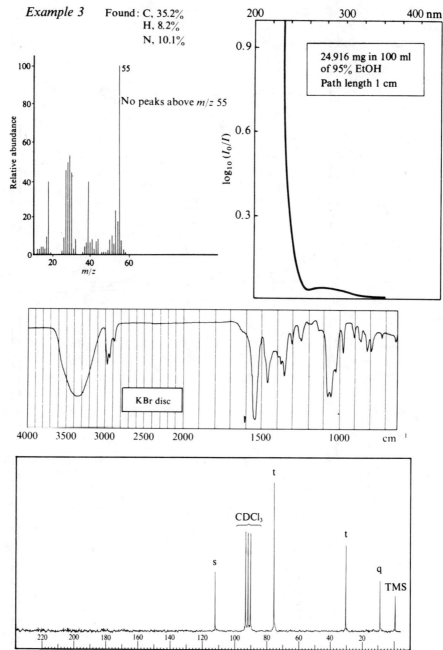

100

No peaks above *m/z* 55

24.916 mg in 100 ml
of 95% EtOH
Path length 1 cm

KBr disc

CDCl₃

TMS

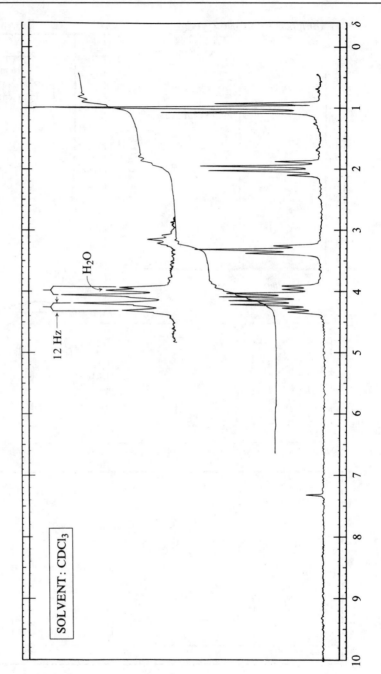

seen in the additional multiplicity ($J = 6$ Hz) in the lower trace. Coupling between CH—OH hydrogens is not always observed (Secs 3.7 and 3.10) but it is here. The resonance at 3.99, which appears in the upper trace, is caused by traces of HOD in the D_2O.

The mass spectrum in this example is unhelpful: there are no striking fragmentations and the only prominent peak, at m/z 55, is misleading, being a $C_4H_7^+$ fragment, which is not found intact in the parent structure.

Example 4

The formula $C_8H_8O_2$ shows that five double bond equivalents are present, and the strong absorption in the UV shows that these must represent considerable unsaturation. With λ_{max} at 316 nm and an ε of 22 000, it is likely that four or five double bonds are conjugated to each other.

The IR spectrum shows that there are few, if any, saturated CHs (only weak bands at just below 3000 cm^{-1}) and only some aryl or unsaturated CHs (the weak band at 3100 cm^{-1}). Since this latter absorption is inherently weak, whereas the former is usually strong, it seems likely that most of the hydrogen atoms are bound to unsaturated centres. The carbonyl region shows two bands: one at 1695 cm^{-1} and the other at 1675 cm^{-1}. These are likely to be $\alpha\beta$-unsaturated ketone, aldehyde, or acid groups, but the last of these is eliminated by the absence of the H-bonded OH absorption in the 3000–2500 cm^{-1} region. An aldehyde group is also ruled out by the absence of absorption in the δ 9–10 region of the ^{1}H NMR spectrum. The compound is, therefore, probably a ketone. The strength and position of the band at 1615 cm^{-1} shows that a C=C conjugated double bond or conjugated aryl group must be present, and bands at 1555 (too weak to be a nitro group) and 1480 cm^{-1} also suggest an aromatic type of compound, although they are untypical enough to indicate that perhaps they are not a simple benzene ring.

The ^{13}C NMR shows eight signals—all, with the exception of a methyl group (the quartet at 27.83), in the unsaturated ($\delta > 100$) region, with a carbonyl carbon at 197.36. In the ^{1}H NMR spectrum, the sharp singlet at δ 2.29 suggests that the methyl group is part of a methyl ketone. This conclusion is supported by the presence in the mass spectrum of the base peak at M–15 (m/z 121). Moreover, a metastable peak at 71.6 establishes the transition m/z 121 $\rightarrow$ m/z 93, and enables us to infer the sequence (**8** $\rightarrow$ **9** $\rightarrow$ **10**). A methyl ketone might also be expected to eliminate ketene (**8** $\rightarrow$ **11**), which this compound does, giving the peak at m/z 94, and to give an acetyl peak (**8** $\rightarrow$ **12**), which is observed (at m/z 43).

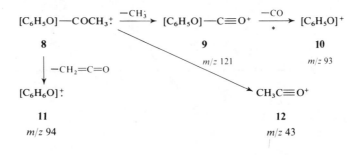

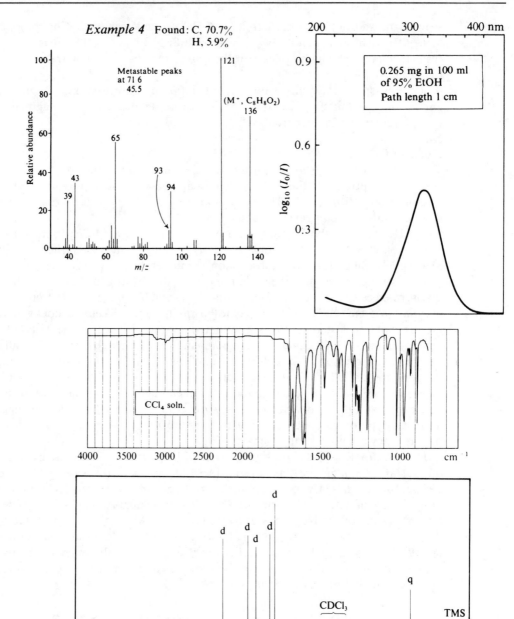

Example 4   Found: C, 70.7%
H, 5.9%

Metastable peaks
at 71 6
45.5

(M⁺, $C_8H_8O_2$)
136

Relative abundance

121

65

43
39

93
94

m/z

0.265 mg in 100 ml
of 95% EtOH
Path length 1 cm

200   300   400 nm

$\log_{10}(I_0/I)$

0.9

0.6

0.3

CCl_4 soln.

4000   3500   3000   2500   2000   1500   1000   cm^{-1}

d

d   d   d
d

q

$CDCl_3$

TMS

s   s

220   200   180   160   140   120   100   80   60   40   20

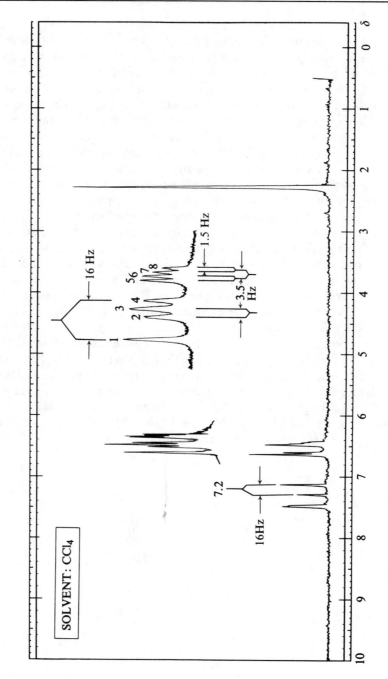

The ^{1}H NMR spectrum also shows that the remaining five hydrogen atoms are attached to double bonds, and the off-resonance ^{13}C NMR shows that each of these are distributed to a different carbon atom (because all the signals are singlets or doublets, after allowing for the methyl quartet at 27.83). The ^{1}H NMR spectrum in CCl$_4$ is unfortunately not well resolved, but the addition of 10 per cent of benzene is enough for us to be able to pick out all the couplings we need to establish the structure. This is shown in the upper trace on the left, which is expanded by a factor of four on the right. In the expanded trace, eight lines can clearly be seen, and in addition there is a doublet at 7.2 and a fine doublet at 7.48, which can be seen on the lower trace. The doublet at 7.2 has an unusually large coupling constant (16 Hz), and it must be coupling to something present in the eight-line pattern. By trial and error we can pick out the lines which are separated by 16 Hz; these turn out to be lines 1 and 4. The large coupling constant and the absence of any further coupling in these lines indicate that we have here an AB system of a *trans* disubstituted (that is, —CH=CH—) double bond. This, incidentally, is supported by the presence in the IR spectrum of a strong band at 970 cm^{-1}, a band typical of this feature (Table 2.3). We have now found fragments **13** and **14** (in which X and Y must carry no protons, since the AB system is not coupled to anything else). The remainder of the molecule is a C$_4$H$_3$O— unit, which does not have a carbonyl group (^{13}C NMR), does not have an OH (IR) and has each of the three hydrogens on a different carbon. With a little thought, the only possibility is seen to be a furan ring with one substituent (**15**), and all we have left to deduce is whether that substituent is on C$_\alpha$ or C$_\beta$. The chemical shifts of the three remaining hydrogen atoms to be assigned (6.46, 6.62, and 7.48) indicate that it is an α-substituted furan (one proton at a low field and two at a high field; Table 3.22). The splittings of the three furan protons amply confirm this assignment: H$_\alpha$ is the fine doublet at 7.48 ($J_{\alpha\beta} = 1.5$ Hz; see Table 3.28), H$_{\beta'}$ is also a doublet (lines 2 and 3, $J_{\beta\beta'} = 3.5$ Hz) and H$_\beta$ is a quartet (lines 5, 6, 7, and 8, $J_{\beta\beta'} = 3.5$ Hz, $J_{\alpha\beta} = 1.5$ Hz).

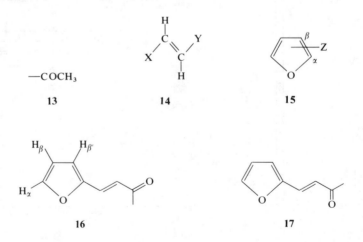

The two carbonyl bands in the IR spectrum are probably due to the presence of the s-*trans* (**16**) and s-*cis* (**17**) conformers. Interconversion between **16** and **17** is rapid at room temperature, and the NMR spectrum therefore presents only a time-averaged picture; it might be possible to see the spectra of both conformers at lower temperatures. The metastable peak at m/z 45.5 in the mass spectrum corresponds to the transition m/z 93 $\rightarrow$ m/z 65, which suggests the elimination of carbon monoxide from **10** to give $C_5H_5^+$.

5.3   Problems

In the following eighteen problems the conditions under which the spectra were obtained are indicated on the actual spectra. Unless otherwise stated, no changes were observed in the NMR spectra after the solutions had been shaken with D_2O. Mass spectra were obtained by electron impact. All metastable peaks that were observed are quoted; some of these, however, may not be useful in interpretation using only the information given in Chapter 4.

The last three problems each include a COSY spectrum and a $^1H/^{13}C$ correlation spectrum, not only to aid the structure elucidation, but also to give practice in reading two-dimensional spectra.

Problem 1 Found: C, 80.7%; H, 7.6%

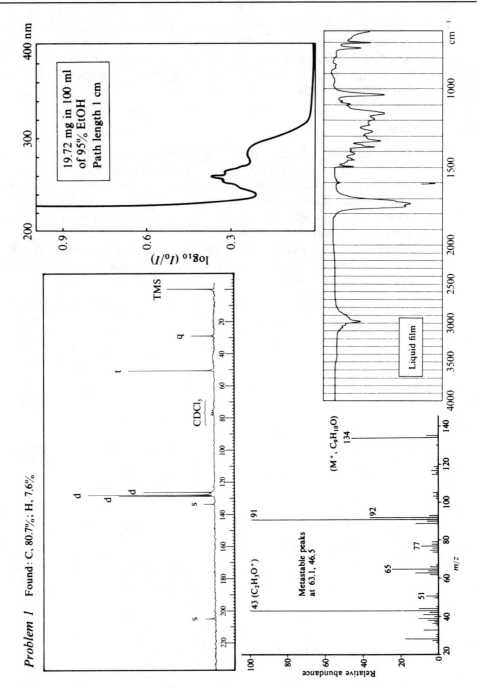

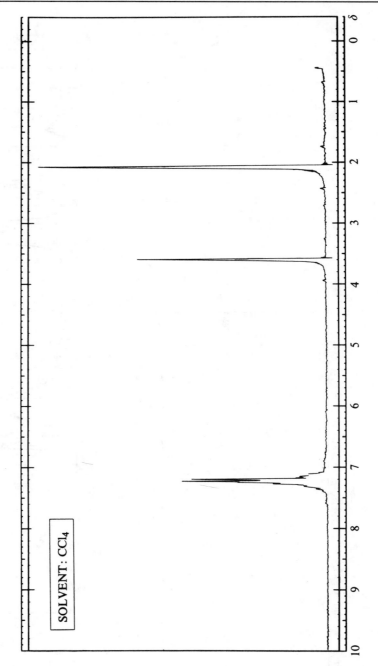

SOLVENT : CCl₄

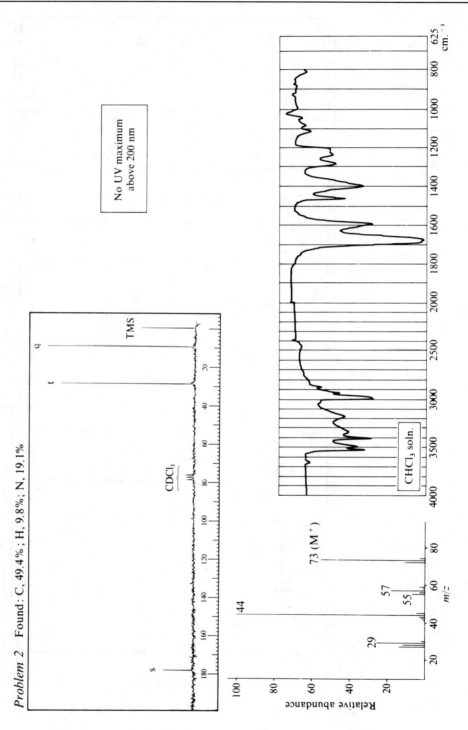

Problem 2   Found: C, 49.4%; H, 9.8%; N, 19.1%

No UV maximum
above 200 nm

CHCl₃ soln.

TMS

CDCl₃

73 (M⁺)

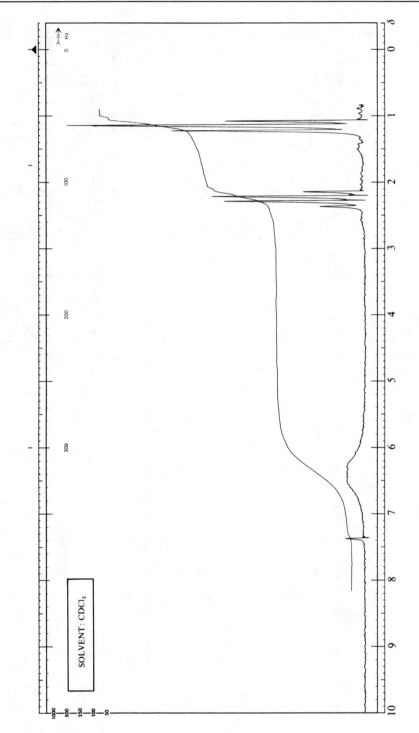

SOLVENT: CDCl₃

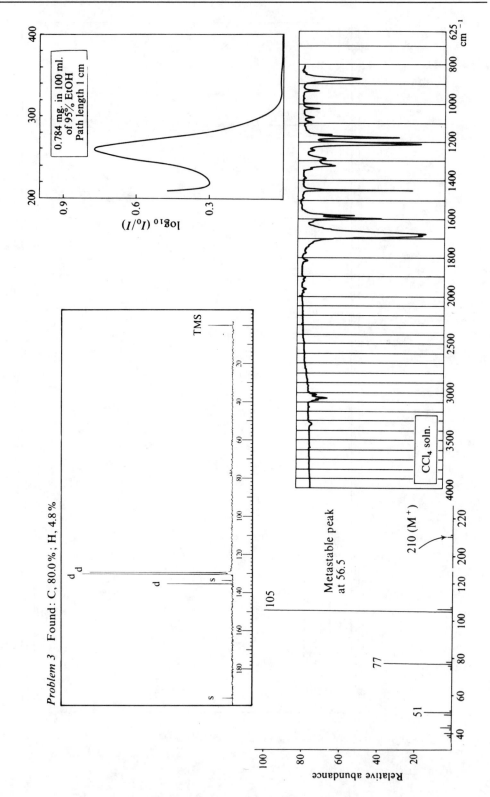

Problem 3   Found: C, 80.0%; H, 4.8%

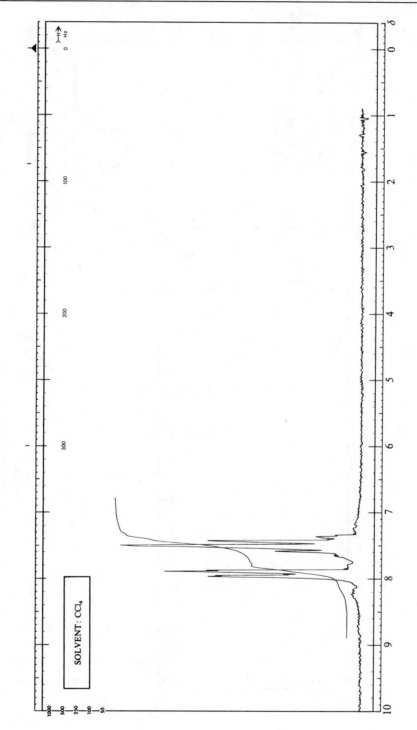

SOLVENT: CCl₄

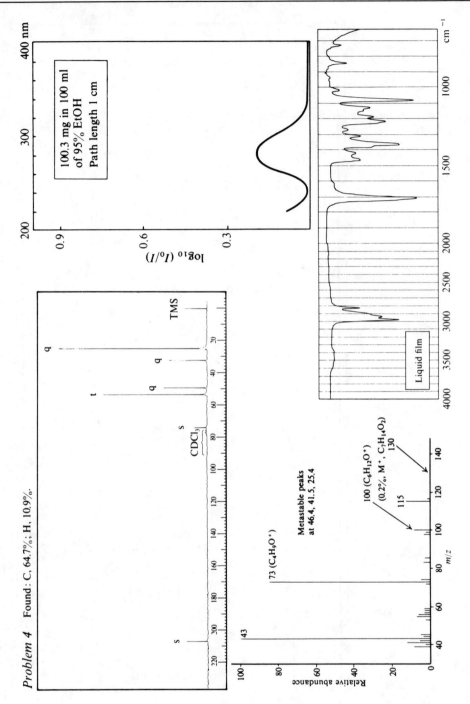

Problem 4   Found: C, 64.7%; H, 10.9%.

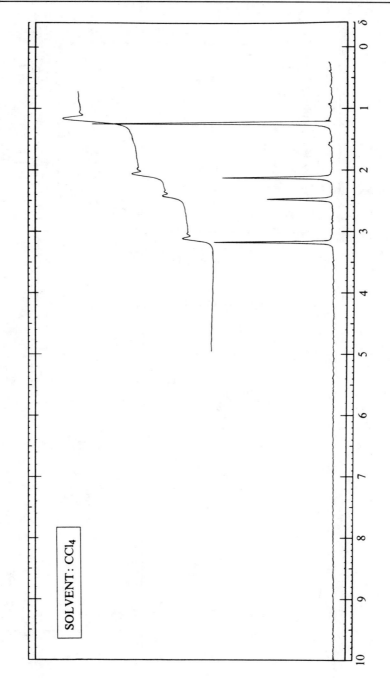

SOLVENT: CCl₄

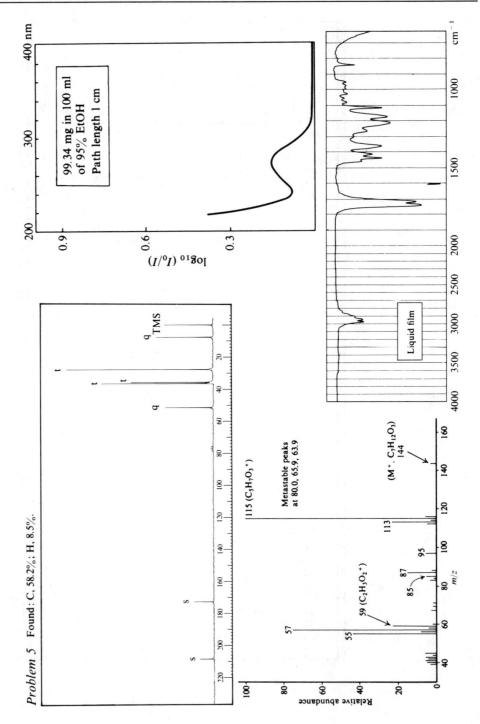

Problem 5   Found: C, 58.2%; H, 8.5%.

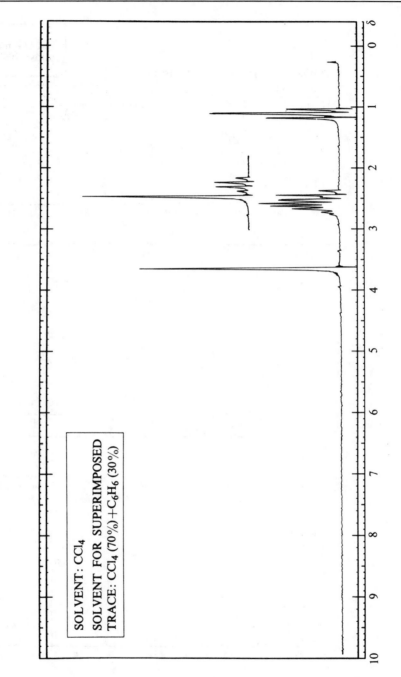

SOLVENT: CCl$_4$
SOLVENT FOR SUPERIMPOSED
TRACE: CCl$_4$ (70%)+C$_6$H$_6$ (30%)

No UV maximum
above 200 nm

Problem 6   Found: C, 36.5%; H, 10.0%.

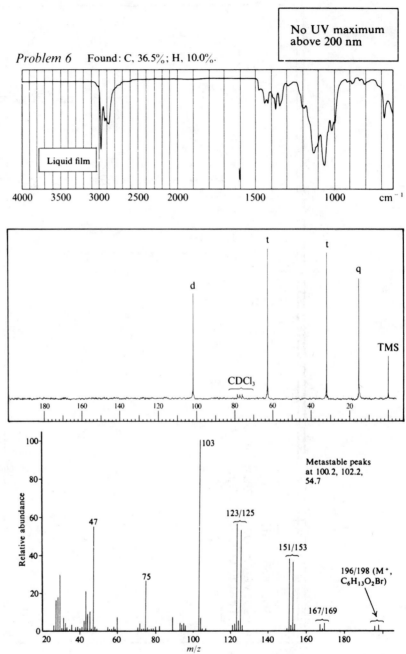

Liquid film

4000   3500   3000   2500   2000   1500   1000   cm^{-1}

d

t

t

q

CDCl$_3$

TMS

180   160   140   120   100   80   60   40   20

Relative abundance

100 — |103

Metastable peaks
at 100.2, 102.2,
54.7

47

123/125

75

151/153

196/198 (M$^+$,
C$_6$H$_{13}$O$_2$Br)

167/169

20   40   60   80   100   120   140   160   180
m/z

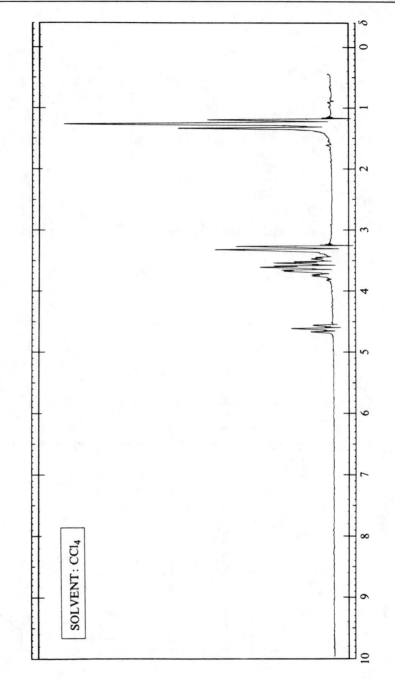

SOLVENT: CCl$_4$

No UV maximum
above 200 nm

Problem 7   Found: C, 39.8%; H, 7.3%

Liquid film

4000   3500   3000   2500   2000   1800   1600   1400   1200   1000   800   625
cm^{-1}

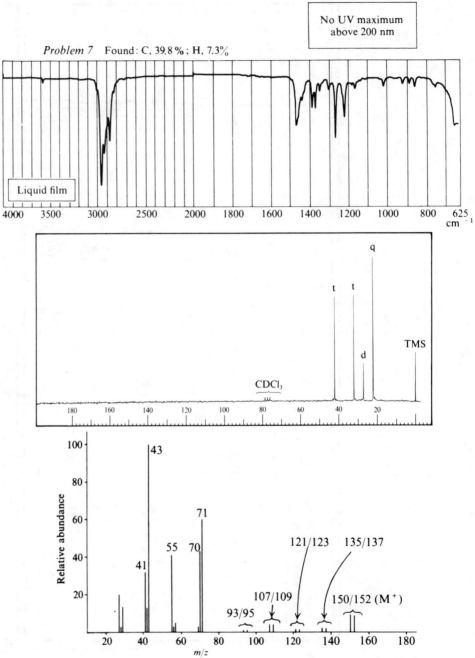

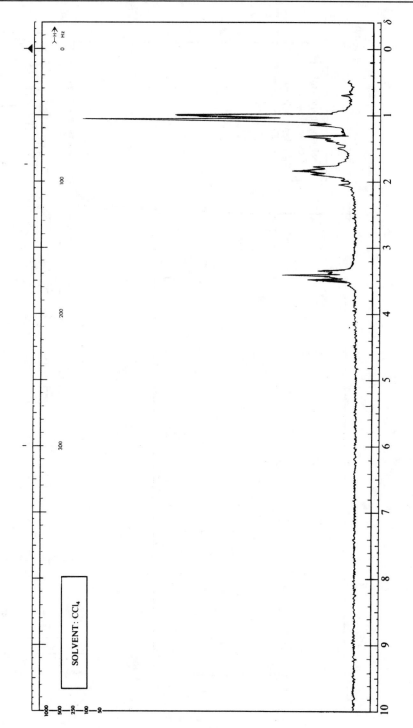

SOLVENT: CCl₄

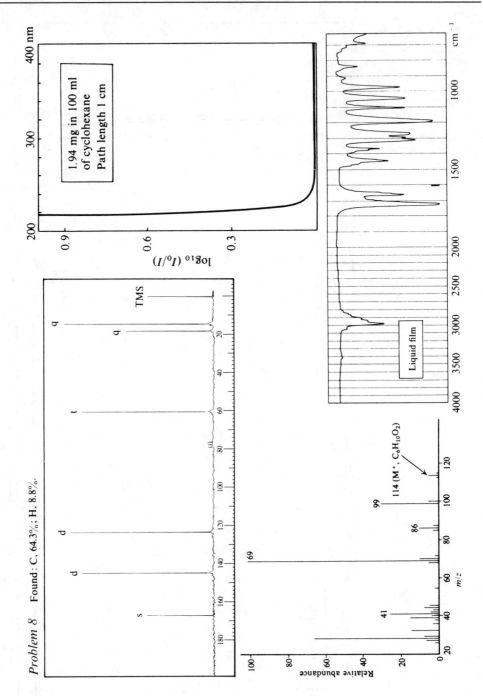

Problem 8   Found: C, 64.3%; H, 8.8%.

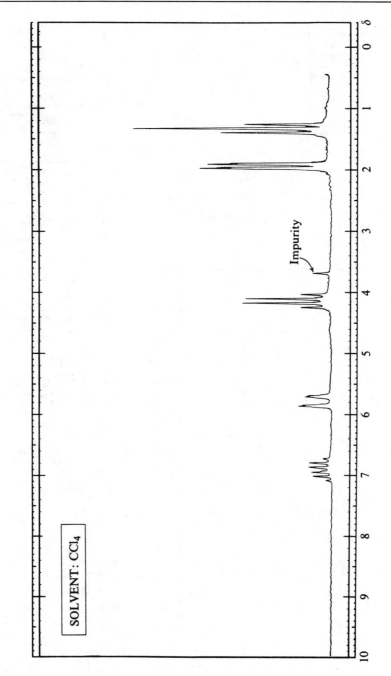

SOLVENT: CCl₄

Impurity

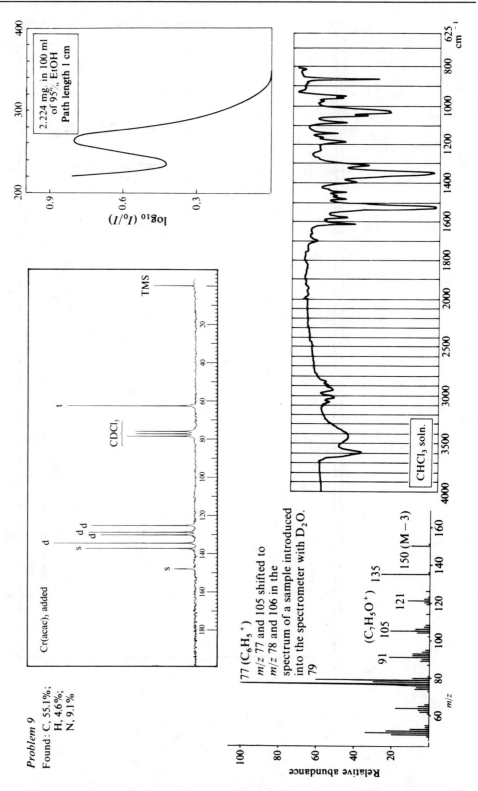

Problem 9
Found: C, 55.1%;
H, 4.6%;
N, 9.1%

2.224 mg. in 100 ml
of 95% EtOH
Path length 1 cm

$\log_{10}(I_0/I)$

Cr(acac)₃ added

TMS

CDCl₃

CHCl₃ soln.

cm⁻¹

77 (C₆H₅⁺)
m/z 77 and 105 shifted to
m/z 78 and 106 in the
spectrum of a sample introduced
into the spectrometer with D₂O.
79

(C₇H₅O⁺) 135
91 105 121   150 (M − 3)

m/z

Relative abundance

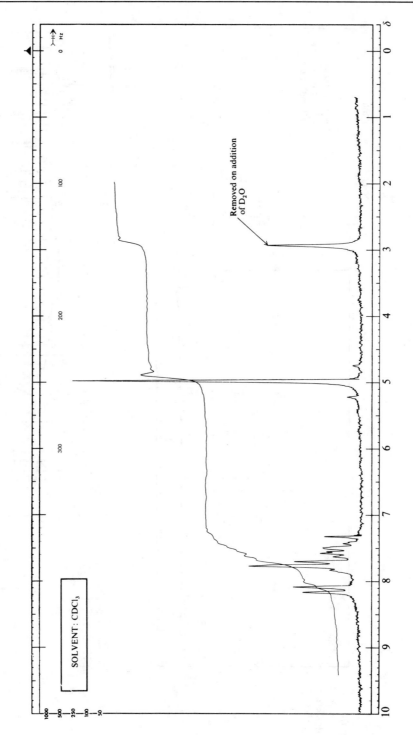

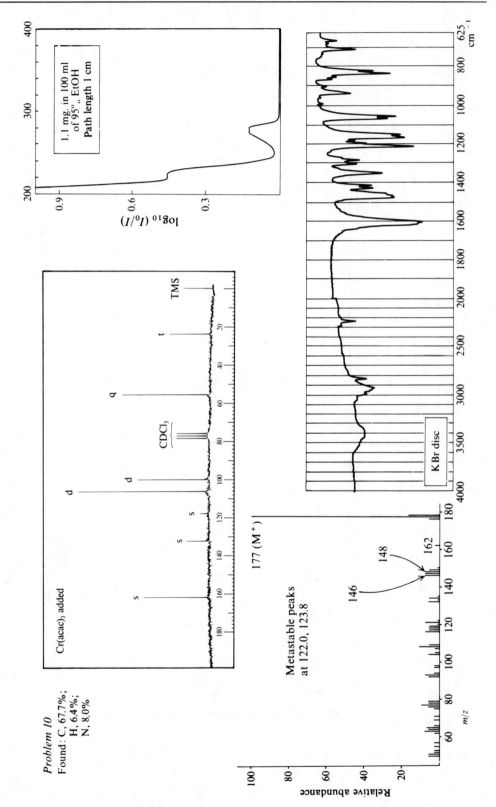

Problem 10
Found: C, 67.7%;
H, 6.4%;
N, 8.0%

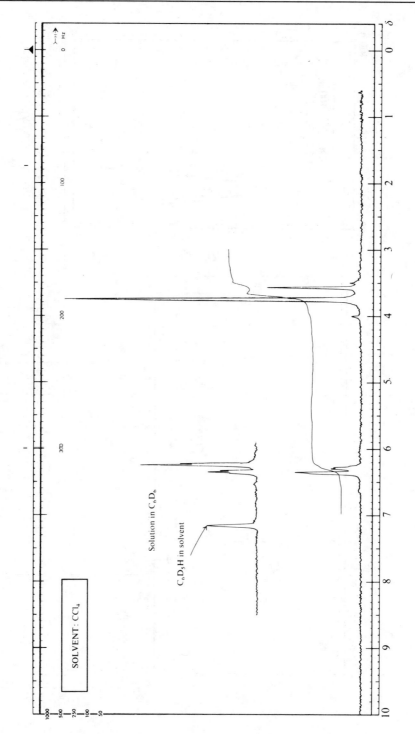

Solution in C_6D_6

C_6D_5H in solvent

SOLVENT : CCl$_4$

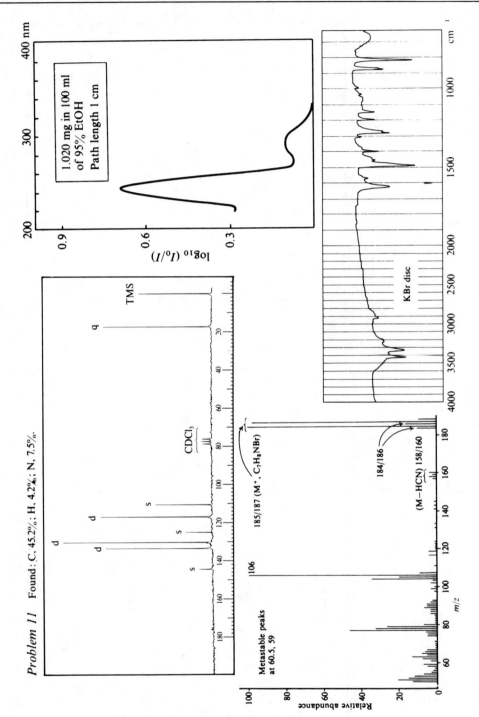

Problem 11   Found: C, 45.2%; H, 4.2%; N, 7.5%.

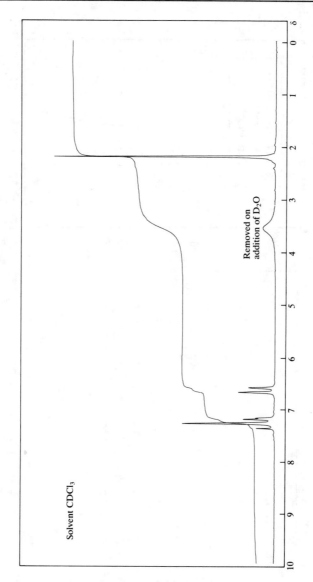

Solvent CDCl$_3$

Removed on addition of D$_2$O

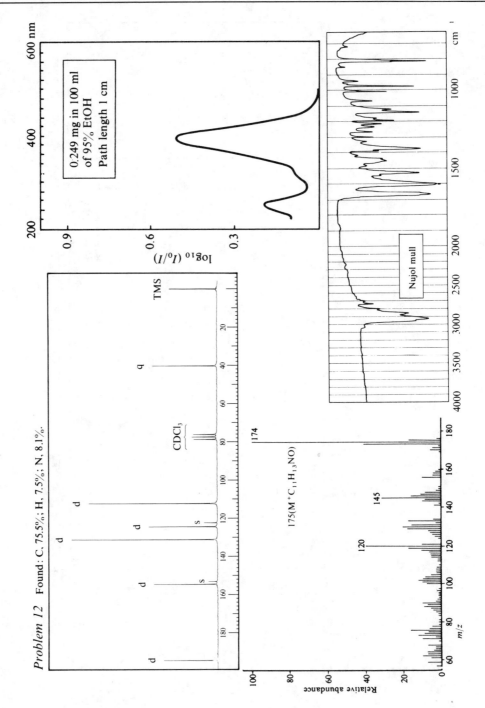

Problem 12   Found: C, 75.5%; H, 7.5%; N, 8.1%.

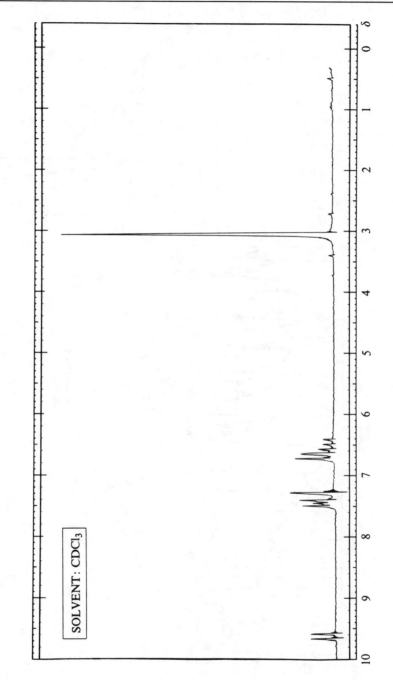

SOLVENT: CDCl$_3$

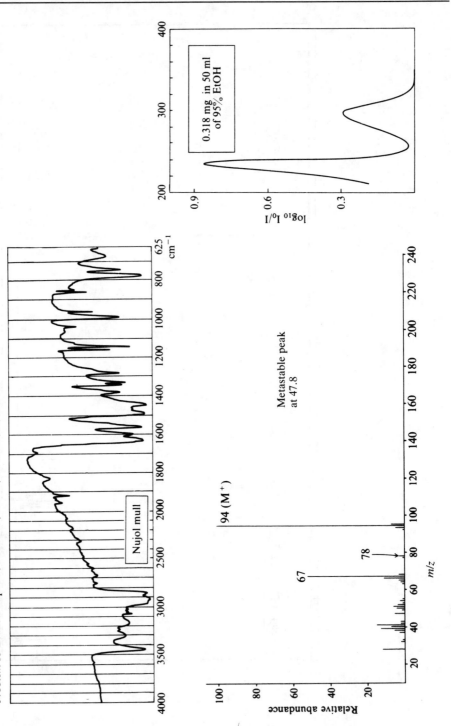

Problem 13 Leaflets m.p. 58°. Found: C, 63.7%; H. 6.5%; N, 29.9%

0.318 mg in 50 ml of 95% EtOH

Nujol mull

94 (M$^+$)

67

78

Metastable peak at 47.8

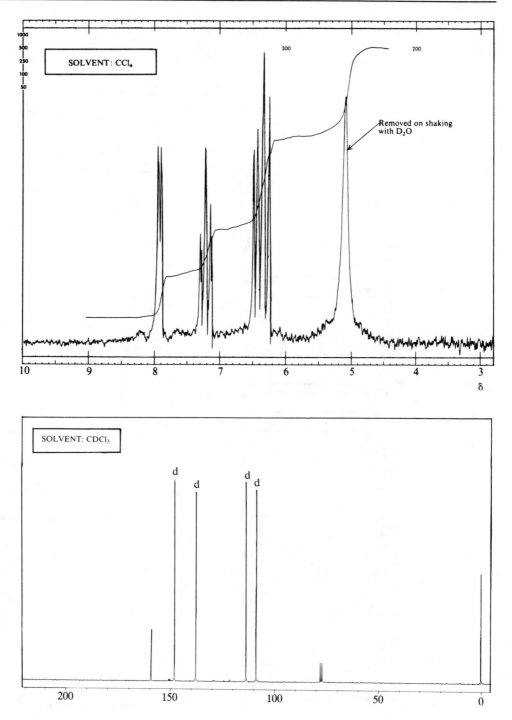

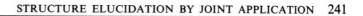

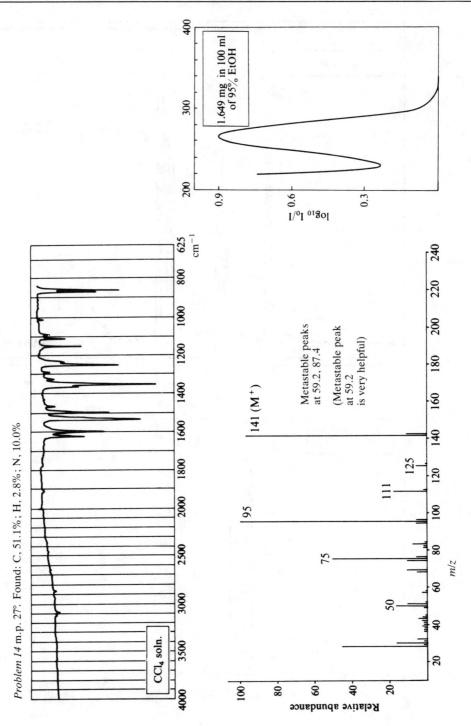

Problem 14 m.p. 27°. Found: C, 51.1%; H, 2.8%; N, 10.0%

CCl₄ soln.

1.649 mg in 100 ml of 95% EtOH

141 (M⁺)

125

111

95

75

50

Metastable peaks at 59.2, 87.4

(Metastable peak at 59.2 is very helpful)

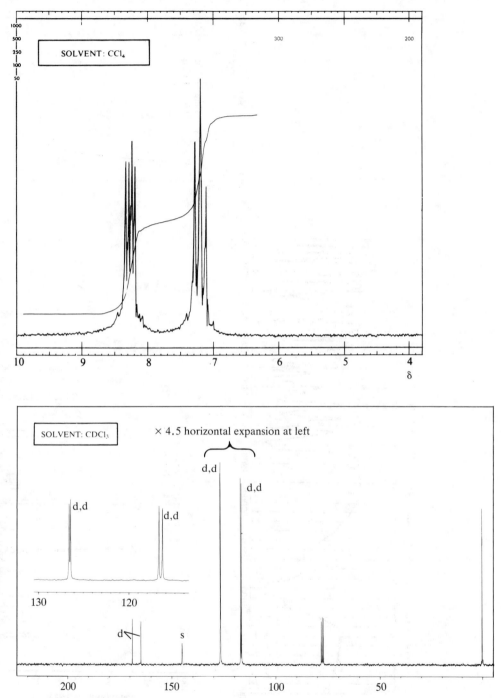

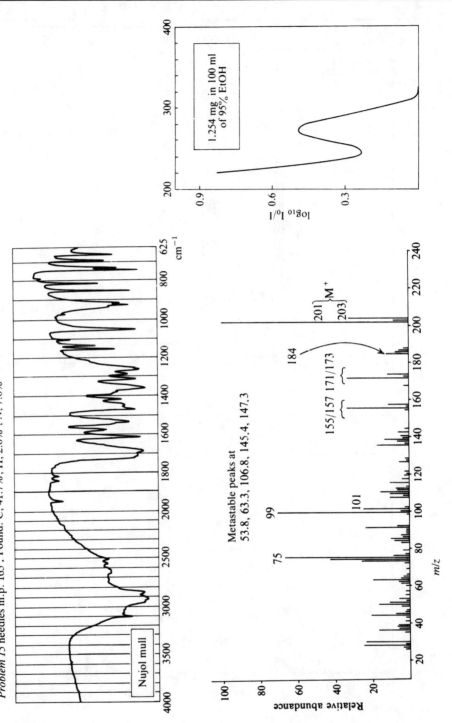

1.254 mg in 100 ml of 95% EtOH

$\log_{10} I_0/I$

cm^{-1}

Nujol mull

Problem 15 needles m.p. 165°. Found: C, 41.7%; H, 2.0%; N, 7.0%

Metastable peaks at 53.8, 63.3, 106.8, 145.4, 147.3

M$^+$

201
203

184

155/157 171/173

101

99

75

m/z

Relative abundance

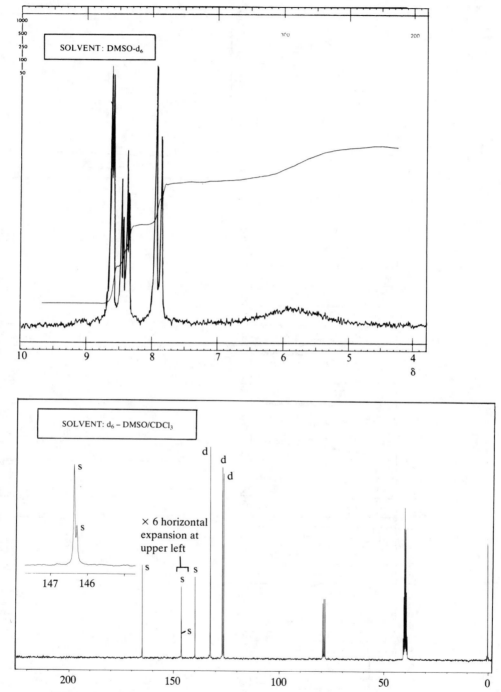

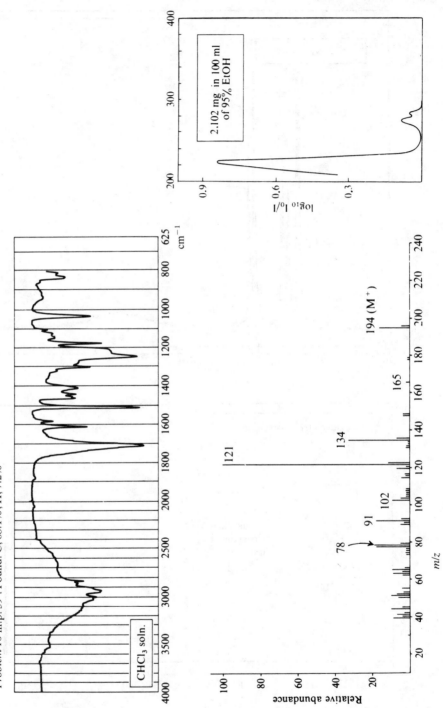

2.102 mg in 100 ml
of 95% EtOH

$\log_{10} I_o/I$

Problem 16 m.p. 59°. Found: C, 68.1%; H, 7.2%

CHCl$_3$ soln.

cm^{-1}

194 (M$^+$)

165

134

121

102

91

78

m/z

Relative abundance

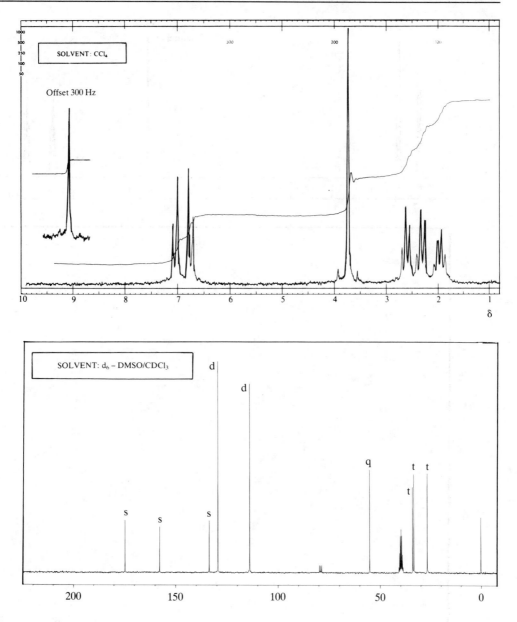

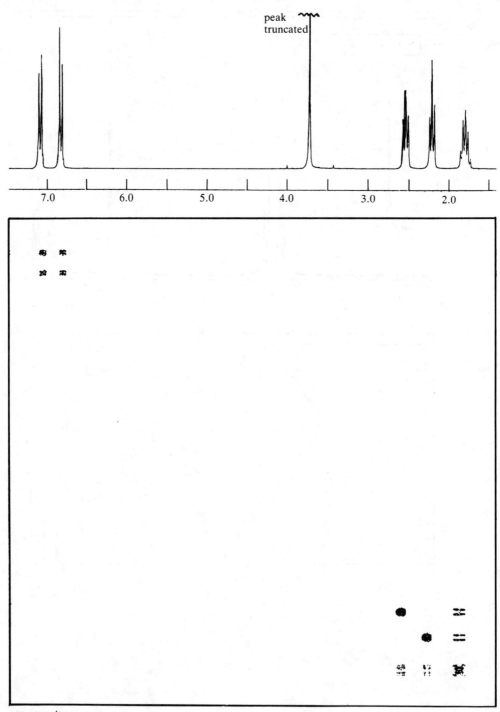

peak
truncated

250 MHz ^{1}H COSY spectrum

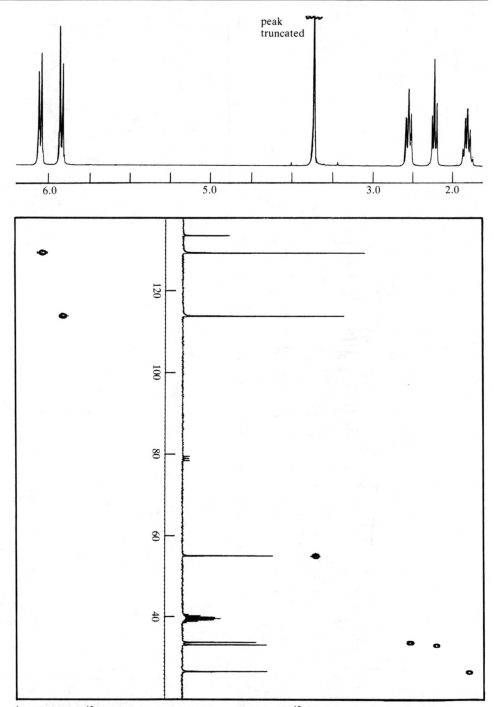

^{1}H (250 MHz)/^{13}C correlation. (The relevant part of the 1D ^{13}C spectrum is displayed within the correlation spectrum to save space; it does not obscure any cross peaks.)

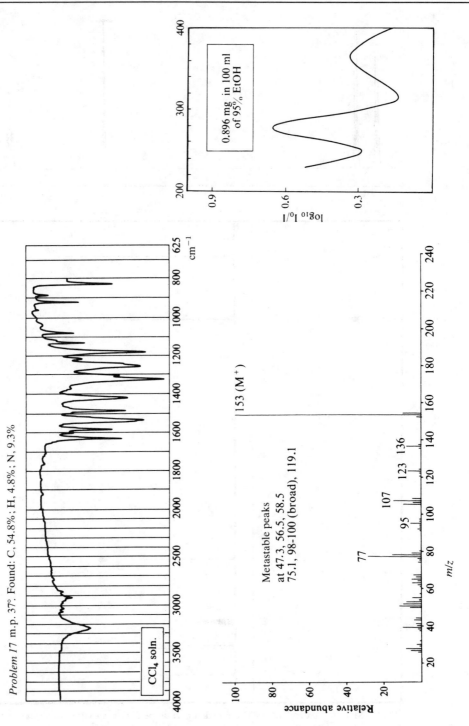

0.896 mg in 100 ml
of 95% EtOH

$\log_{10} I_0/I$

Problem 17 m.p. 37°. Found: C, 54.8%; H, 4.8%; N, 9.3%

CCl$_4$ soln.

cm^{-1}

153 (M$^+$)

Metastable peaks
at 47.3, 56.5, 58.5
75.1, 98-100 (broad), 119.1

Relative abundance

m/z

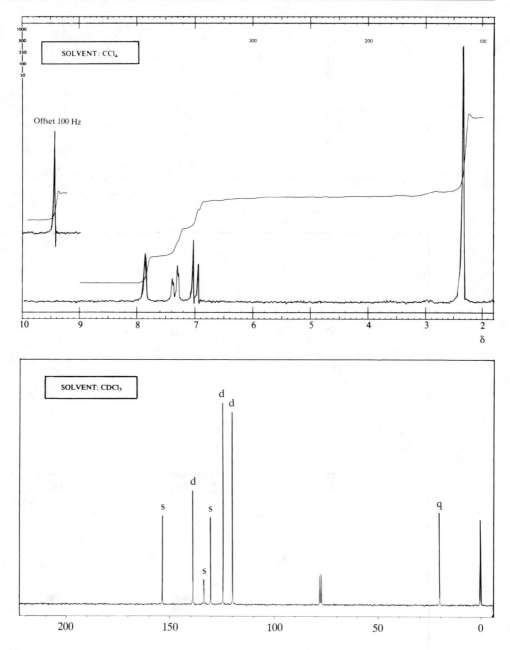

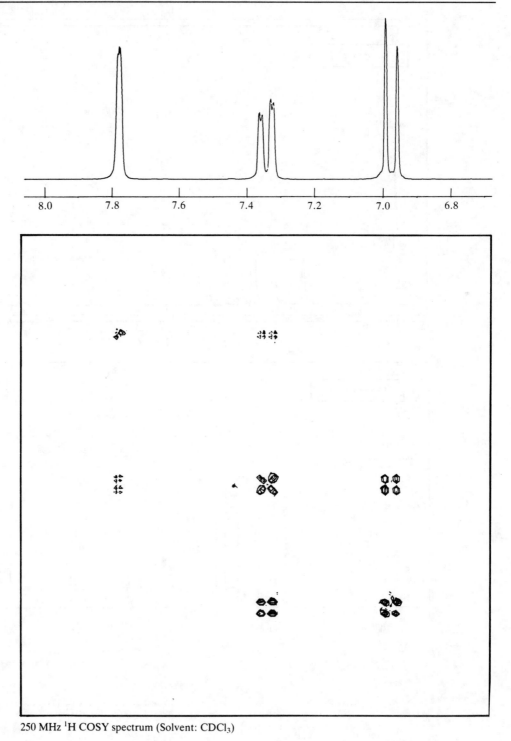

250 MHz ^{1}H COSY spectrum (Solvent: CDCl$_3$)

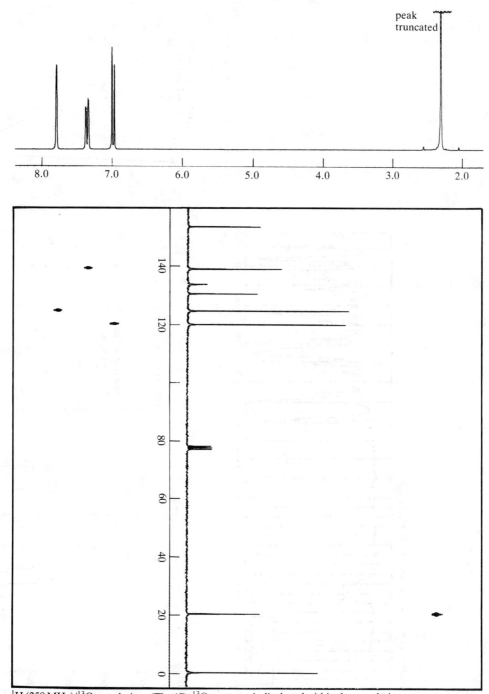

peak
truncated

^{1}H (250 MHz)/^{13}C correlation. (The 1D ^{13}C spectrum is displayed within the correlation spectrum to save space, it does not obscure any cross peaks.)

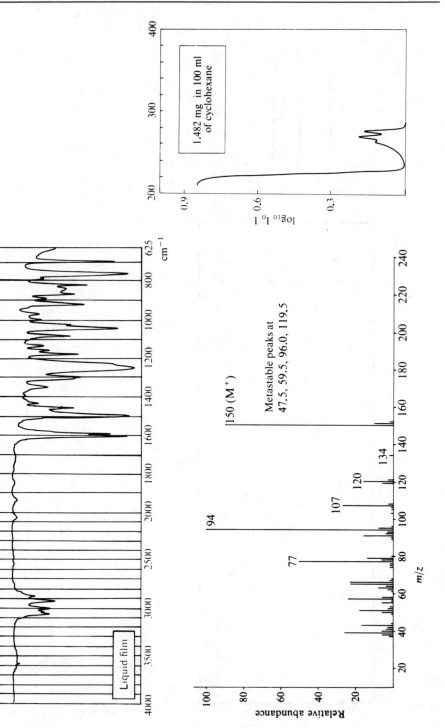

1.482 mg in 100 ml of cyclohexane

$\log_{10} \dfrac{I_0}{I}$

cm^{-1}

Problem 18 b.p. 243°, Found: C, 71.8%; H, 6.8%

Liquid film

150 (M⁺)

Metastable peaks at 47.5, 59.5, 96.0, 119.5

94

77

107

120

134

Relative abundance

m/z

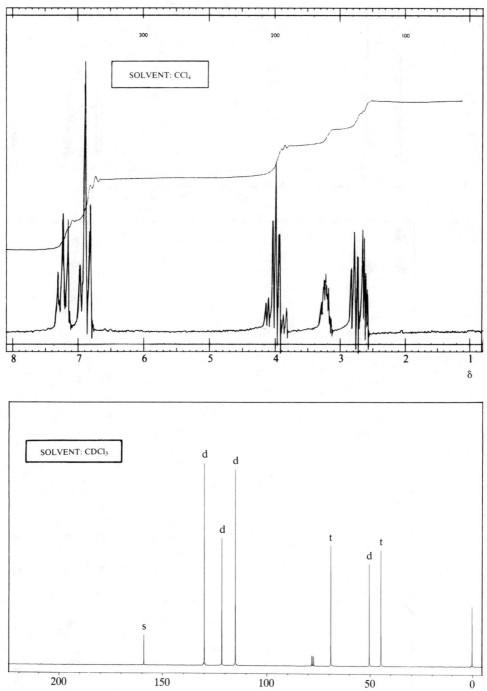

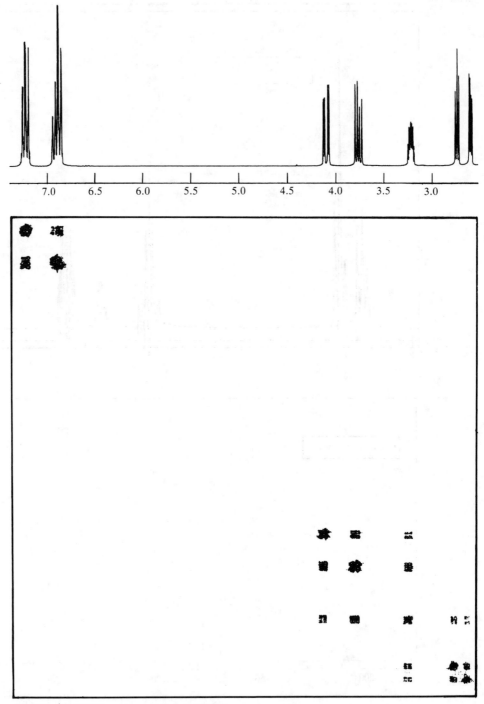

250 MHz ^{1}H COSY spectrum (Solvent: CDCl$_3$)

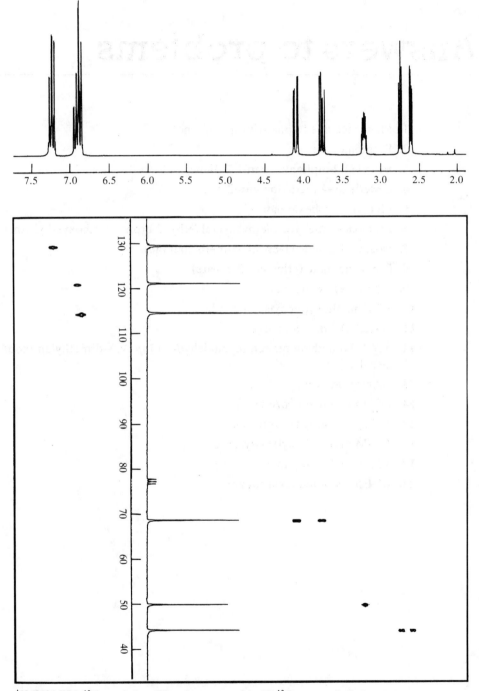

^{1}H (250 MHz)/^{13}C correlation. (The relevant part of the 1D ^{13}C spectrum is displayed within the correlation to save space; it does not obscure any cross peaks.)

Answers to problems

1. Phenylacetone (1-phenylpropan-2-one)
2. Propionamide
3. Benzil (1,2-diphenylethan-1,2-dione)
4. 4-Methoxy-4-methylpentan-2-one
5. Methyl 4-oxohexanoate
6. Bromoacetaldehyde diethyl acetal (ethyl 2-bromo-1-ethoxyethyl ether)
7. Isoamyl bromide (1-bromo-3-methylbutane)
8. Ethyl crotonate (ethyl but-2-enoate)
9. *o*-Nitrobenzyl alcohol
10. 3,5-Dimethoxyphenylacetonitrile
11. 2-Methyl-4-bromoaniline
12. *p-N,N*-Dimethylaminocinnamaldehyde   [3-(*p-N,N*-dimethylaminophenyl)-prop-2-en-1-al]
13. 2-Aminopyridine
14. 1-Fluoro-4-nitrobenzene
15. 2-Chloro-5-nitrobenzoic acid
16. 4-(*p*-Methoxyphenyl)-butyric acid
17. 4-Methyl-2-nitrophenol
18. 1,2-Epoxy-3-phenoxypropane

Index

Absorbance, UV, 2
Alkenes:
    IR, 41, 51–52, 59, 60, 62, 211, 214
    NMR, 72, 97, 99, 129–130, 137–138
    Mass, 171
Allylic coupling, 98, 99, 101
Amino acids, masses of, 186
Anisotropy, NMR, 71, 74
[18]-Annulene:
    UV, 26
    NMR, 72
Aromatic compounds:
    UV, 18–25
    IR, 41, 46, 53, 58, 59, 60, 61, 211
    NMR, 72, 131–132, 139–140
    Mass, 172–175
Auxochrome, definition of, 8

Bathochromic effect, definition of, 7
Beer's Law, 2
Benzene:
    UV, 18–19
    NMR, 70, 72
    as solvent in NMR, 75
Blue shift, 8, 28
Boltzmann distribution, 63

Carbonyl group:
    UV, 4, 5, 6, 17, 201
    IR, 38, 47–51, 57, 58, 59, 60, 61, 62, 201, 204, 211, 214
    NMR, 71, 133, 137, 201, 207, 211
    Mass, 176, 177
Carboxylic acids:
    UV, 18
    IR, 42–43, 49, 58
    NMR, 75, 141
    Mass, 177
Chemical ionization, 153–155, 182–184
Chemical shift:
    definition of, 66–67
    factors affecting, 69–76
    Tables of values
        ^{13}C, 126–133
        ^{1}H, 135–142

Chromophore, UV:
    definition of, 4–7
    non-conjugated, 27
Coalescence, NMR, 102, 103
Conjugation, importance in UV, 5
COSY spectra, 112–114
Cotton effect, 17
Coupling, *see* spin-spin coupling
Coupling constants:
    definition of, 76
    factors affecting magnitude of, 91–101
    in aromatic compounds, 95, 99, 134
    in ring systems, 93–95, 99, 134
    Tables of values
        ^{13}C-^{19}F, 134
        ^{13}C-^{1}H, 134
        ^{19}F-^{1}H, 147
        ^{1}H-^{1}H, 143–146
        ^{31}P-^{1}H, 148
Cross peaks, NMR, 113, 119, 122, 123
CW, 64, 65
Cycloheptadiene, UV, 11
Cyclohexadiene:
    UV, 11
    NMR, 99, 139
Cyclopentadiene:
    UV, 11
    NMR, 139
Cyclopropanes:
    IR, 35, 40
    NMR, 74, 126, 134, 136

Data systems, Mass, 189–190
Decoupling, NMR, 76, 79, 80, 110–114
δ, definition of, 67, 68
Deshielding:
    convention for direction of, 68
    definition of, 70
Deuterium exchange, NMR, 75, 83
Dienes:
    UV, 8–11
    NMR, 129, 139, 144
Difference spectroscopy, NMR:
    decoupling, 111–112
    NOE, 116–118
Double bond equivalents, 199–200

Double resonance, NMR, 110–114
Downfield, convention for direction of, 68

Electron impact (EI) ionization, 152
Energetics of fragmentation, Mass, 165–172
Exchange phenomena, NMR, 83
Extinction coefficient, UV, 2, 200–201

FAB mass spectra, 155–157
Field desorption, 155, 184
Field, high or low, definition and convention for, 68
Field strength, NMR, 64
Fingerprint:
    UV, 23
    IR, 33, 34, 39, 54–57
Fourier transform:
    IR, 31
    NMR, 63, 66, 69, 106
Fragmentation, in mass spectra, 165–186
Frank-Condon principle, 5
Free induction decay, 65
FT, see Fourier transform
Functional group determination by MS, 185

Gas chromatography-mass spectrometry (GC/MS), 187–188
Geminal coupling:
    definition of, 85
    factors affecting, 96–98
    Tables of values, 143

Heats of formation of ions, 167–172
Heterocyclic compounds:
    UV, 24–25
    IR, 62
    NMR, 131, 134, 136, 139, 214
    Mass, 175
Homoallylic coupling, 98–101, 146
Hydrogen bonding:
    UV, 6
    IR, 31, 42–44, 48, 49, 50, 51, 52, 57, 58
    NMR, 75, 141
Hyperchromic effect, definition of, 8
Hypochromic effect, definition of, 8
Hypsochromic effect, definition of, 8

Infrared spectra:
    acetates, 40, 56
    acetylenes, 45, 58

acid halides, 47
alcohols, 42, 57, 208
aldehydes, 40, 48, 60, 61
alkenes, 41, 51–52, 59, 60, 62, 211, 214
amides, 43, 49–50, 59–61
amines, 43–44
amino acids, 43, 49, 59
anhydrides, 47
aromatic compounds, 41, 46, 53, 58, 59, 60–61
boron compounds, 56
carbonates, 51
carbonyl groups, 38, 47–51, 57, 58, 59, 60, 61, 62, 201, 204, 211, 214
carboxylic acids, 42–43, 49, 58
C-H stretching, 32–33, 35, 40–41, 57, 58, 59, 60, 61, 62
cumulated double bonds, 36, 45–46
cyclopropanes, 35, 40
deuterium substitution, 44
double bonds, 37–38, 46, 47–54
epoxides, 35, 40, 54
esters, 47–48, 61, 204
ethers, 54
fingerprint region, 33–34, 39, 55–57, 61
functional groups, correlation charts of, 35–39
halogen compounds, 56
heterocyclic compounds, 62
imides, 50
imines, 52
inorganic ions, 57
intensities, 34, 51
ketones, 40, 48–49, 57, 201, 211, 214
lactams, 43, 50
lactones, 47, 62
measurement of, 29–31
N-H bending, 37, 44
N-H stretching, 35, 43, 59, 60–61
nitriles, 45, 62
nitro compounds, 54, 58, 208
nitroso compounds, 54
N-oxides, 54
Nujol, 30, 34
O-H stretching, 42–43, 57, 58, 208
olefins, see alkenes
oximes, 52
phenols, 42, 48, 49
phosphorus compounds, 56
ring strain, 51, 52–53
selection rules, 32
single bonds to hydrogen, 35, 40–44
solvents for, 30
silicon compounds, 55
sulphur compounds, 55–56
triple bonds, 36, 45, 58, 62

αβ-unsaturated carbonyl compounds, 47–50
vibrational modes, 33
'Inside' hydrogens, NMR, 72
'Inside' lines, NMR, 91
Integration, NMR, 69, 201
Intensity of absorption:
  UV, 6
  IR, 34
  NMR, 69
Ion analysis, 158–164
Ion cyclotron resonance, 163–164
Isosbestic point, UV, 8
Isotope abundances and masses, 164–165
Isotope labelling:
  IR, 44
  Mass, 192

Karplus equations, 92

Lambert's law, 2
Line broadening, NMR, 101–103
Liquid chromatography–mass spectrometry (LC/
  MS), 188–189
Long-range coupling, NMR, 96, 98–101, 113, 146

Magnetic sector instruments, 158–162
Mass spectra:
  acetals, 176, 179–180
  alcohols, 176
  aldehydes, 177
  aliphatic compounds, 175–180
  alkanes, 177
  alkenes, 171
  amides, 177
  amines, 176–178
  aromatic compounds, 168, 172–175
  bromine, isotope abundance of, 164–165
  ^{13}C, abundance of, 164–165
  californium plasma desorption, 157–158
  carbonium ions, 171
  carboxylic acids, 177
  chlorine, isotope abundance of, 164–165
  competing reactions, 150
  computer systems, 189–191
  esters, 169–170, 173, 175–177, 182–183, 207
  ethers, 176
  ethylene ketals, 176, 179–180
  exact masses of common atoms, 165
  fast atom bombardment (FAB), 155–157
  gas chromatography systems, 187–188
  high resolution, 159
  γ-hydrogen rearrangement, 170, 177, 207
  impurity peaks, 180
  interpreting an unknown, 193
  ion cyclotron resonance (ICR), 163–164
  ionization, 152–158
  isotopic labelling, 192
  ketals, see ethylene ketals
  ketones, 176–178
  laser desorption, 156
  liquid chromatography systems, 188–189
  matrix, in FAB, 156
  measurement of, 150
  metastable peaks, 166–167
  molecular ion, 152–158
  nitro compounds, 175, 181
  ortho effect, 175
  peptide sequencing, 185–187
  polyfunctional molecules, 176, 179–180
  rearrangement processes, 169–171
  representation of, 150–151
  resonance effects, 175
  sample size, 158, 185
  secondary ion mass spectrometry (SIMS), 155
  spectrometers, 158–164
  sulphur, isotope abundance of, 165
  Table of m/z values, 194–197
  Table of common losses, 181
  time-of-flight, 163
Mass spectrometry–mass spectrometry (MS/MS),
  160–162, 184
Measurement of spectra:
  UV, 2
  IR, 29–31
  NMR, 66
  Mass, 150, 158–164
meta coupling, NMR, 99, 113
Metastable peaks, 166–167
Methylene envelope, NMR:
  definition of, 103
  resolving signal in, 106, 108–110, 111–112, 118
Molecular ion, 152–158
Multiple ion monitoring (MIM), 191–193

Norbornadiene, UV, 27
Nuclear Magnetic Resonance:
  AA', BB' systems, 91
  AB systems, 85–88, 90, 95, 214
  ABX systems, 88–90
  aldehydes, 71, 133, 137
  alkenes, 72, 97, 99, 129–130, 137–138
  allenes, 99, 143, 146
  allylic coupling, 98–101, 146
  angle strain, effect of, on coupling constants, 95
  anisotropy, 71, 74

Nuclear Magnetic Resonance: *contd.*
  aromatic compounds
    chemical shifts in, 72, 131–132, 139–140
    coupling constants in, 99, 134, 145–147
  benzene
    anistropy of, 70, 72
    as solvent, 75, 214
  benzenes, substituted, 131–132, 139–140
  bond lengths, effect on coupling constants, 95
  carbonyl compounds, 71, 73, 96, 133, 137, 148,
    201, 207, 211
  chemical shift
    definition of, 66–67
    factors affecting, 69–76, 106
    Tables of values:
      ^{13}C, 126–133
      ^{1}H, 135–142
  coalescence, 102, 103
  contour plots
    COSY, 113
    NOESY, 120
    ^{13}C-^{1}H correlated, 122
    INADEQUATE, 123
  correlated spectra, *see* contour plots
  COSY, 112–114
  coupling, *see* spin–spin coupling
  coupling constants
    definition of, 76
    factors affecting magnitude of, 91–101
    in aromatic compounds, 95, 99, 134
    in ring systems, 93–95, 99, 134, 214
    Tables of values:
      ^{13}C-^{19}F, 134
      ^{13}C-^{1}H, 134
      ^{19}F-^{1}H, 147
      ^{1}H-^{1}H, 143–146
      ^{31}P-^{1}H, 148
  cyclohexanes, 92–94, 99, 126, 136
  cyclopropanes, 74, 126, 134, 136
  D$_2$O shake, *see* deuterium exchange
  decoupling, *see* spin decoupling
  δ, definition of, 67, 68
  deshielding effect, 68
  deuterated solvents, 66, 142
  deuterium exchange, 75, 83, 208
  diastereotopic groups, definition of, 88
  difference decoupling spectra, 111–112
  dihedral angle, effect of, on coupling constants,
    91–93
  double bond, anistropy of, 71
  downfield, convention for direction of, 68
  electronegative and electropositive elements
    effect on chemical shift, 70–72
    effect on coupling constants, 93–94

epoxides, 94
ethyl groups, 83, 201, 204, 208
exchange phenomena, 101–103
^{19}F-^{1}H coupling constants, *see* coupling
    constants
Fourier transform, 63, 65, 66, 106
geminal coupling:
  definition of, 85
  factors affecting, 96–98, 143
heterocyclic compounds, 131, 134, 136, 139, 214
high-field spectra, 64, 65, 105–106
homoallylic coupling, 98–101, 146
hydrogen bonding, 75, 141
inductive effect, on chemical shift, 69–70
integration, 69, 201
intensity, 69
Karplus equations, 92
line broadening, 101–103
line perturbation, 86, 87
long-range coupling, 98–101, 113, 146
measurement of, 66
meta coupling, 99, 113
methylene envelope:
  definition of, 103
  resolving signals in, 106, 108–110, 111–112,
    118
NH signals, 75, 83, 141, 148
NOESY spectra, 119–120
nuclear Overhauser enhancement:
  origin of, 114–116
  NOE difference spectra, 116–118
  NOESY spectra, 119–120
off-resonance decoupling, 79, 80
OH signals, 75, 83, 102, 141, 148, 208
^{31}P-^{1}H coupling constants, *see* coupling
    constants
π-contribution to geminal coupling, 96–97
'pointing' of signals in AB systems, 86, 101
proton noise decoupling, 76
quadrupole relaxation, 83, 102
relaxation, 64–69, 101, 114
resolution, 67, 68
ring current, 70, 72, 74
sample size, 66
SH signals, 75, 83, 141
shielding effect, 68
shift reagents, 103–105
solvents, 66, 75, 83, 142
spin decoupling
  COSY, 112–114
  difference, 111–112
  heteronuclear and homonuclear, defined, 110
  off resonance, 79
  simple, 76, 110–111

spin–spin coupling (*see also* coupling constants, Tables of)
^{13}C-^{13}C, 80
^{13}C-^{1}H, 76, 80
^{13}C-^{2}H, 76
first order rules, 78–79, 81–90
geminal, 85, 96–98, 143
^{1}H-^{1}H:
    first-order, 81–85
    deviations from first-order, 90–91
    separating on to different axis from δ, 106–110
    simple splitting patterns, 85–89
^{1}H-^{2}H, 96, 142
long-range, 98–101, 113, 146
mechanism of, 91, 96, 99
vicinal:
    definition of, 81
    factors affecting, 91–96, 144–145
W, 99, 146
stacked plots, 108
temperature effect, 75, 102–103
TMS, 66, 67, 69
αβ-unsaturated carbonyl compounds, 71–72, 137–138
upfield, convention for direction of, 68
Nuclear Overhauser effect:
    origin of, 114–116
    NOE difference spectra, 116–118
    NOESY spectra, 119–120
Nuclear spin, 63, 65, 114, 125
Nuclei, magnetic data on, 125
Nujol, 31, 34, 59, 61

Off-resonance decoupling, 79–80
'Outside' hydrogens, NMR, 72

Paramagnetic salts, use of, in NMR, 69, 76, 101, 103–105
Peptide sequencing, 185–187
Plasma desorption, Mass, 155–158
Pulsed spectra, NMR, 64, 106

Quadrupole mass spectrometers, 162–163
Quadrupole relaxation, NMR, 83, 102

Radicals, 167–172
Raman spectra, 32, 45, 49, 51, 52, 55, 62
Red shift, 7, 27, 28

Relaxation, NMR, 64, 69, 101, 114
Restricted rotation, 103
Ring current, 70, 72, 74

Sample size:
    UV, 2
    IR, 30–31
    NMR, 66
    Mass, 158, 185
Selection rules:
    UV, 3–4, 22
    IR, 32
Shielding, NMR:
    convention for direction of, 68
    definition of, 70
Shift reagents, 103–105
Single ion monitoring (SIM), 191
Solvents:
    UV, 3, 5–6, 15–16, 17
    IR, 30, 61
    NMR, 66, 75, 83, 142
Specific ion monitoring, 190–193
Spin decoupling:
    COSY, 112–114
    difference, 111–112
    heteronuclear and homonuclear, defined, 110
    off resonance, 79
    simple, 76, 110–111
Spin–spin coupling:
    ^{13}C-^{13}C, 80
    ^{13}C-^{1}H, 76, 80
    ^{13}C-^{2}H, 76
    first-order rules, 78–79, 81–90
    geminal, 85, 96–98, 143
    ^{1}H-^{1}H
        first-order, 81–85
        first-order, deviations from, 90–91
        separating on to different axis from chemical shift, 106–110
        simple splitting patterns, 85–89
    ^{1}H-^{2}H, 96, 142
    long-range, 98–101, 113, 146
    mechanism of, 91, 96, 99
    vicinal
        definition of, 81
        factors affecting, 91–96, 144–145
    W, 99, 147
Steric effects, UV, 26, 27–28

τ, definition of, 135
Tautomerism, 25
Time-of-flight mass spectrometers, 163

Ultraviolet spectra:
  aldehydes, 15–17, 22
  benzene, 18
  benzenes, substituted, 19–22
  chromophores, 3–5, 8
    non-conjugated, 4–5, 17, 27, 28
  dienes, 5, 8–11
  α-diketones, 17
  fine structure of, 2–3, 19
  heteroaromatic compounds, 23–25, 28
  ketones, 4–5, 6, 14–18, 20, 22, 201
  measurement of, 2
  $n \rightarrow \pi^*$ transition, 4, 6, 17, 26, 27, 201, 208
  $\pi \rightarrow \pi^*$ transition, 5–6, 8–17, 18–27
  polycyclic aromatic hydrocarbons, 23
  polyenes, 11–12
  polyenynes, 13–14
  poly-ynes, 13–14
  porphyrins, 26
  quinones, 26
  reference works, 7, 28
  sample size, 2
  selection rules for, 3–4
  solvent effects, 5–6, 15–16, 17
  solvents for, 3
  $\alpha\beta$-unsaturated
    acids, 18
    aldehydes, 14–17
    amides, 18
    esters, 18
    ketones, 14–17
    nitriles, 18
    Woodward's rules, 9, 15
Units:
  UV, 2
  IR, 29
  NMR, 67, 76, 135
Upfield, convention for direction of, 68

Vicinal coupling:
  definition of, 81
  factors affecting, 91–96, 144–145
Visible spectra, 1–28

Woodward's rules, 9, 15